GROUP THEORY AND HOPF ALGEBRAS
Lectures for Physicists

T0324955

GROUP THEORY AND HOPF ALGEBRAS

Lectures for Physicists

A P Balachandran
Syracuse University, USA

S G Jo
Kyungpook National University, Korea

G Marmo
University Federico II Naples, Italy

 World Scientific

NEW JERSEY · LONDON · SINGAPORE · BEIJING · SHANGHAI · HONG KONG · TAIPEI · CHENNAI

Published by

World Scientific Publishing Co. Pte. Ltd.

5 Toh Tuck Link, Singapore 596224

USA office: 27 Warren Street, Suite 401-402, Hackensack, NJ 07601

UK office: 57 Shelton Street, Covent Garden, London WC2H 9HE

Library of Congress Cataloging-in-Publication Data
Balachandran, A. P., 1938–
    Group theory and Hopf algebras : lectures for physicists / by A.P. Balachandran,
S.G. Jo & G. Marmo.
      p. cm.
    Includes bibliographical references and index.
    ISBN-13: 978-981-4322-20-1 (hardcover : alk. paper)
    ISBN-10: 981-4322-20-2 (hardcover : alk. paper)
    1. Group theory. 2. Hopf algebras. I. Jo, S. G. II. Marmo, Giuseppe. III. Title.
QC20.7.G76B35 2010
512'.2--dc22

                                                                          2010018280

British Library Cataloguing-in-Publication Data
A catalogue record for this book is available from the British Library.

Printed in Singapore.

Preface

For several years, one of the authors (A. P. B.) taught a one-semester graduate course in group theory to second-year graduate students in physics at Syracuse University. The students in this course were expected to have a prior knowledge of the familiar tools of mathematical physics such as linear algebra and complex analysis and a good background in classical and quantum mechanics. These students were preparing to engage in such diverse fields of research as condensed matter and particle physics, and the course was designed to give them a broad culture in group theory.

In 1984, Balachandran and Trahern published a book entitled "Lectures on Group Theory for Physicists". It was based on these courses and addressed to second-year graduate students in physics with a reasonable understanding of linear vector spaces, classical mechanics and quantum mechanics who seeks a general education in group theory and to teachers of such a course as well. Classroom experience showed that the material of this book can be covered without difficulty in one semester. This was an elementary book for physicists so that no attempt was made at rigor. Furthermore, since it was not addressed to students in one or another particular field, there was also a conscious attempt to treat the subject from a broad perspective and to avoid lengthy discourses on specialized topics like finite groups or Lie algebras.

Recently, Balachandran and his colleagues Sang Jo and Giuseppe Marmo decided to revise this book. One reason for our decision has been the emergent significance of Hopf algebras or quantum symmetries in fundamental physics. Symmetry groups provide simple, but particular examples of Hopf algebras. The latter, however, are more general and yet retain the features required of a symmetry in quantum theory. Quantum symmetries can thus be based on Hopf algebras which do not come from groups. Hopf algebras such as $SU(2)_q$ with features transcending those of groups first made their appearance in physics during the study of integrable models and conformal field theories. Recently, they have also assumed a central role in quantum field theories on noncommutative spacetimes. But there is still no widespread appreciation of the far-reaching importance of Hopf algebras for fundamental physics. It seemed to us that a revision of the 1984 book which includes a simple introduction to Hopf algebras is now appropriate.

On the other hand, we did not want to make the unduly long. Our intention has been to write a book which can be used for a one-semester course like the old one, or for a two-semester course which also treats relatively advanced topics.

The book is divided into five parts and still maintains the lecture note style. The first three essentially comprise the material of the old version. In part IV, besides relatively minor additions and explanations of the old version, we have now included a discussion of the Galilei group and its relation to the Poincaré group via group contractions. We have also included a discussion of the reduction of the direct product representations of the Poincaré group and the derivation of the Landau-Yang theorem on the non-existence of the $Z^0 \to 2\gamma$ from there. This theorem is a striking group theoretical result which does not rely on quantum field theory at all.

Part V contains an introductory treatment of Hopf algebras. The literature on Hopf algebras and their applications is extensive. Our intention is not to give an in-depth review of this subject. Rather we want to explain why it is progressively recognized that quantum symmetries need not be based on groups and instead can be based on Hopf algebras and their variants, and then introduce the reader to their elementary properties. We wish also to illustrate the applications of these algebras in physics using examples from quantum field theories on noncommutative spacetimes.

There are other beautiful topics of interest to physicists in Lie group theory which we do not discuss here. There is for example a whole body of literature which discuss coadjoint orbits of the Lie groups. They are symplectic manifolds, and in favorable cases, all unitary representations of the parent group can be obtained by quantizing them. We have discussed this material in detail elsewhere.[1,2] It can be used to supplement this part.

---

[1] A. P. Balachandran, G. Marmo, B.-S. Skagerstam and A. Stern, *Gauge Symmetries and Fibre Bundles* (Springer-Verlag, Berlin, 1983).

[2] A. P. Balachandran, G. Marmo, B.-S. Skagerstam and A. Stern, *Classical Topology and Quantum States* (World Scientific, Singapore, 1991).

There is also a sophisticated literature on Kac-Moody and Virasoro algebras which has a fundamental role in conformal field theories and string physics. We refer the reader to specialized treatises for their treatment.[3,4]

The first three parts of the book, like its old version, should be accessible to a second year student working towards a doctoral degree on a topic in theoretical physics. Parts IV and V are somewhat more demanding and require greater mathematical sophistication. Part V on Hopf algebras in particular contains introductory material to active research areas in physics and mathematics.

We thank C. G. Trahern for generously granting us permission to use the material of the old book in this new version, and Francesco del Franco and Bibliopolis for waiving copyright. This book would have been impossible to write without their cooperation.

---

[3]See for example P. D. Francesco, P. Mathieu and D. Sénéchal, *Conformal Field Theory* (Springer-Verlag, Berlin, 1997).

[4]See for example M. Green, J. H. Schwarz and E. Witten, *Superstring Theory*, volumes 1 and 2 (Cambridge University Press, 1987).

Contents

THE POINCARÉ GROUP   **155**

HOPF ALGEBRAS IN PHYSICS   193

PART 1
GENERAL NOTIONS

LECTURE 1

i) Definition of a group and examples

Definition. A group G is a set $S = \{g\}$ such that for any two elements $g_1, g_2 \in S$, a composition law $\circ$ called a product is defined. It has the properties:

1) $g_1, g_2 \in G \Longrightarrow g_1 \circ g_2 \in G$. We write $g_1 \circ g_2 = g_1 g_2$.
2) The product is associative : $g_1(g_2 g_3) = (g_1 g_2)g_3$.
3) G contains an identity e. That is, $ge = eg = g, \forall g \in G$.

It follows that the identity is unique. For if e' is another identity, then $ee' = e = e'$. e is sometimes written as 1, and if the composition law is addition (+), then it is written as 0.

4) For every $g \in G$ there is an inverse, $g^{-1} \in G$, such that $gg^{-1} = g^{-1}g = e$.

The inverse is unique. For if $e = g^{-1}g = \bar{g}g$, multiply on the right by $g^{-1} \Rightarrow g^{-1} = \bar{g}$.

Definition. The order of a group is the number of elements in it. It can be finite, countably infinite, or uncountably infinite.

Examples:

1) The permutation group S_n on n objects. The order is $n!$.
Take n objects, number them 1, 2, ..., n. Put them in n boxes in a

3

row. For $n = 4$, we could have :

2	1	3	4

A permutation $s \in S_n$, written as

$$s = \begin{pmatrix} 1 & 2 & \dots & n \\ s_1 & s_2 & \dots & s_n \end{pmatrix}$$

sends object k to object s_k. Thus if

$$s = \begin{pmatrix} 1 & 2 & 3 & 4 \\ 4 & 1 & 3 & 2 \end{pmatrix},$$

then

$$s \;\boxed{2}\,\boxed{1}\,\boxed{3}\,\boxed{4} = \boxed{1}\,\boxed{4}\,\boxed{3}\,\boxed{2}\,.$$

The order of the columns of s is immaterial.

The group arises here from maps of an underlying set (the set being objects in boxes). It is a 'transformation group'. We now verify the group properties:

Closure :

$$\begin{pmatrix} 1 & 2 & \dots & n \\ t_1 & t_2 & \dots & t_n \end{pmatrix} \begin{pmatrix} 1 & 2 & \dots & n \\ s_1 & s_2 & \dots & s_n \end{pmatrix} = \begin{pmatrix} 1 & 2 & \dots & n \\ t_{s_1} & t_{s_2} & \dots & t_{s_n} \end{pmatrix} \in S_n.$$

Associativity :

$$\begin{pmatrix} 1 & 2 & \dots & n \\ u_1 & u_2 & \dots & u_n \end{pmatrix} \left[\begin{pmatrix} 1 & 2 & \dots & n \\ t_1 & t_2 & \dots & t_n \end{pmatrix} \begin{pmatrix} 1 & 2 & \dots & n \\ s_1 & s_2 & \dots & s_n \end{pmatrix} \right]$$

$$= \begin{pmatrix} 1 & \dots & n \\ u_1 & \dots & u_n \end{pmatrix} \begin{pmatrix} 1 & \dots & n \\ t_{s_1} & \dots & t_{s_n} \end{pmatrix} = \begin{pmatrix} 1 & \dots & n \\ u_{t_{s_1}} & \dots & u_{t_{s_n}} \end{pmatrix}$$

$$\left[\begin{pmatrix} 1 & 2 & \dots & n \\ u_1 & u_2 & \dots & u_n \end{pmatrix} \begin{pmatrix} 1 & 2 & \dots & n \\ t_1 & t_2 & \dots & t_n \end{pmatrix} \right] \begin{pmatrix} 1 & 2 & \dots & n \\ s_1 & s_2 & \dots & s_n \end{pmatrix}$$

$$= \begin{pmatrix} 1 & \dots & n \\ u_{t_1} & \dots & u_{t_n} \end{pmatrix} \begin{pmatrix} 1 & \dots & n \\ s_1 & \dots & s_n \end{pmatrix} = \begin{pmatrix} 1 & \dots & n \\ u_{t_{s_1}} & \dots & u_{t_{s_n}} \end{pmatrix}.$$

Identity:

$$e = \begin{pmatrix} 1 & \dots & n \\ 1 & \dots & n \end{pmatrix}.$$

Inverse:

$$\begin{pmatrix} 1 \dots n \\ s_1 \dots s_n \end{pmatrix}^{-1} = \begin{pmatrix} s_1 \dots s_n \\ 1 \dots n \end{pmatrix}.$$

S_n can also be thought of as linear transformations on a complex n-dimensional vector space $\mathbb{C}^n$. Let $e_1, e_2, \dots, e_n$ be a basis of this vector space. Define a family of $n!$ linear operators by

$$L\left[\begin{pmatrix} 1 & 2 \dots & n \\ s_1 & s_2 \dots & s_n \end{pmatrix} \right] \sum_i \xi_i e_i = \sum_i \xi_i e_{s_i}$$

where $\sum_i \xi_i e_i$ is a generic vector of $\mathbb{C}^n$. If s, t are permutations, then one sees easily that $L(s)L(t) = L(st)$. The set $\{L(s)\}$ is clearly a group with group multiplication as operator multiplication. It can also be thought of as S_n. [See Lecture 2.]

S_n appears in the discussion of i) Pauli exclusion principle in quantum mechanics and ii) some continuous groups and all finite groups.

2) The rotation group $O(3)$ in 3 dimensions.

Consider $\mathbb{R}^3 =$ real three-dimensional vector space, i.e., $\mathbb{R}^3 = \{x = (x_1, x_2, x_3) \,|\, x_i \text{ real}\}$. The linear transformations $x \to Rx$, $x_i \to R_{ij}x_j$, which preserve the scalar product $(x, y) = \sum x_i y_i$ are such that $R^T R = 1$ (hence R is real and orthogonal), and $\det R^2 = 1$ or $\det R = 1$ or -1.

The set of all real, orthogonal matrices with $\det = +1$ is the group $SO(3)$ of rotations *without* inversions. ("Det" is an abbreviation for determinant.) The set of all rotations (including reflections) is the group $O(3)$. The set of all elements with $\det = -1$ does not form a group. (There is no identity, and the set is not closed under multiplication.)

Let

$$P = \begin{bmatrix} -1 & & \\ & -1 & \\ & & -1 \end{bmatrix} \in O(3)$$

be "parity". We have $\det P = -1$ and $P^2 = e$. If $R \in O(3)$ had $\det R = -1$, write $R = PR'$ (where $R' = PR$). Here $R' \in O(3)$ and $\det R' = +1 \Rightarrow R' \in SO(3) \Rightarrow O(3) = SO(3) \cup PSO(3)$.

The corresponding groups in n dimensions are $SO(n)$, $O(n)$.

3) The group $SU(2) = \{g\}$. Here g is a 2×2 unitary matrix with $\det g = +1$. $SU(2)$ leaves $(z, z') = \sum_{i=1}^{2} z_i^* z_i'$ invariant. A similar definition holds for $SU(n)$ while $U(n) = \{g\}$, g being any $n \times n$ unitary matrix.

4) The Lorentz group $\mathcal{L} = \{\Lambda\}$. Λ is a real 4×4 matrix which leaves the form $(x, y) = \sum x_\mu y^\mu = x_0 y_0 - \sum_{i=1}^3 x_i y_i$ invariant. If

$$\eta = \begin{bmatrix} 1 & & & 0 \\ & -1 & & \\ & & -1 & \\ 0 & & & -1 \end{bmatrix},$$

then

$$(x, y) = x_i \eta_{ij} y_j$$

and

$$\Lambda^T g \Lambda = g.$$

5) The Poincaré group $\mathcal{P}$ consists of all Lorentz transformations and translations. An element of $\mathcal{P}$ is (a, Λ) where $a = (a_0, a_1, a_2, a_4) \in \mathbb{R}^4$ and $\Lambda \in \mathcal{L}$. On a four-vector x,

$$(a, \Lambda)x = \Lambda x + a,$$

i.e. first Lorentz transform, and then translate. [This choice of order is only a convention.]

Now

$$\begin{aligned}
(a', \Lambda')(a, \Lambda)x &= (a', \Lambda')(\Lambda x + a) \\
&= (\Lambda'(\Lambda x + a) + a') \\
&= (\Lambda' a + a', \Lambda' \Lambda)x
\end{aligned}$$

or

$$(a', \Lambda')(a, \Lambda) = (a' + \Lambda' a, \Lambda' \Lambda).$$

The identity is $(0, 1)$ and the inverse of (a, Λ) is

$$(a, \Lambda)^{-1} = (-\Lambda^{-1} a, \Lambda^{-1}).$$

The groups above appear concretely as *transformation groups* on underlying sets. In the associated *abstract* group, we just consider the group structure (and forget about their origin as transformation groups).

LECTURE 2

i) Mapping and functions for sets

Consider two sets E and F, which can be either distinct or the same.

A law f which assigns to each $x \in E$, a unique element $f(x) \in F$ is a function from E to F.

If to each $y \in F$, there is an $x \in E$ such that $y = f(x)$, then the map f is onto F.

If the image of E under f is not all of F, the map is into F.

If to each $y \in F$, $\exists$ only one $x \in E$ such that $f(x) = y$, the map is one-to-one (1-1). In this case,

$$f(x_2) = f(x_1) \Rightarrow x_2 = x_1.$$

The symbol f^{-1} is defined by: for $y \in F$, $f^{-1}(y)$ is the *set* in E such that each member of the set is mapped to y by f. f^{-1} is in general not a function unless f is 1-1.

ii) Isomorphisms and homomorphisms

Assume that the mapping is onto. For if the mapping were not onto, we can restrict attention to the image.

1) A map f of a group G onto a group G' is a homomorphism if it preserves the group laws, i.e.

$$f(g_1)f(g_2) = f(g_1 g_2).$$

This rule implies the following :

(a) If $e \in G$, $e' \in G'$ are identities then $f(e) = e'$. For $f(e)f(g) = f(eg) = f(ge) = f(g)f(e) = f(g)$. Since f is an onto map, and e' is unique, $f(e) = e'$.

(b) $f(g^{-1}) = f(g)^{-1}$. For $f(g^{-1})f(g) = f(g^{-1}g) = e'$.

The set of all elements in G mapped to e' by f is called the *kernel* of the homomorphism f and denoted as Ker f.

2) If in 1), f is 1-1, then the homomorphism is an *isomorphism*. Then G and G' as groups are identified.

3) If $G' = G$, then replace "homo" by "endo", and "iso" by "auto" in the above.

Examples:

1) Let

$$s = \begin{pmatrix} 1 \dots n \\ s_1 \dots s_n \end{pmatrix} \in S_n.$$

Let $L(s)$ be the linear operator on $\mathbb{C}^n$ with basis $e_1, e_2, \dots, e_n$ defined by

$$L(s)e_i = e_{s_i}.$$

Then $\{L(s)\}$ forms a group under multiplication, $L(s)L(t) = L(st)$. Call this group $S_n' \equiv \{L(s)\}_{s \in S_n}$. Then

$$L : S_n \longrightarrow S_n', \ s \xrightarrow{L} L(s)$$

is an isomorphism.

2) Let g_0 be a fixed element in a group G. Consider

$$G \ni g \xrightarrow{f_{g_0}} g_0 g g_0^{-1} \in G.$$

This map is an automorphism, a so-called *inner* automorphism.

Proof :

$$1-1 : g_0 g g_0^{-1} = g_0 g' g_0^{-1} \Rightarrow g = g'$$

Onto : For arbitrary $g \in G$, $g_0^{-1} g g_0 \in G$ and this is mapped by f_{g_0} to g.

Homomorphism: $f_{g_0}(g_1)f_{g_0}(g_2) = (g_0 g_1 g_0^{-1})(g_0 g_2 g_0^{-1}) = g_0(g_1 g_2)g_0^{-1} = f_{g_0}(g_1 g_2)$.

iii) $SU(2)$ and $SO(3)$

There is a 2-1 homomorphism R from $SU(2)$ onto $SO(3)$ with $\pm U \in SU(2)$ having the same image $R(U) \in SO(3)$. $SU(2)$ is called the *covering group* (in fact the *universal covering group*) of $SO(3)$.

We show the following [and hence the existence of the 2-1 homomorphism]:

1. There exists an R which maps U to a real 3×3 matrix $R(U)$ such that

$$R(U_1)R(U_2) = R(U_1 U_2).$$

2. The map is into $SO(3)$. [i.e. $R(U)^T R(U) = 1$, $\det R(U) = +1$.]
3. The map is onto $SO(3)$.
4. $R(U_1) = R(U_2) \Rightarrow U_1 = \pm U_2$.

Proofs:

1) Let τ_1, τ_2, τ_3 be Pauli matrices : $\tau_i^\dagger = \tau_i$, $\mathrm{tr}\ \tau_i = 0$. Any 2×2 matrix m is a linear combination of the unit matrix 1 and τ_i:

$$m = \alpha_0\, 1 + \alpha_i\, \tau_i.$$

If m is traceless and $m^\dagger = m$, then

$$\alpha_0 = 0$$

and

$$\alpha_i \tau_i = \alpha_i^* \tau_i \ \ or \ \ \alpha_i^* = \alpha_i.$$

Thus any traceless hermitian 2×2 matrix is a real linear combination of the τ_i's.

Now consider $U\, \tau_i\, U^\dagger = \tau_i'$ with $U \in SU(2)$. Then

$$Tr\ \tau_i' = 0, \ \ \tau_i'^\dagger = \tau_i' \Rightarrow U\, \tau_i\, U^\dagger = R_{ji}(U)\tau_j$$

where the real $R_{ji}(U)$ are uniquely defined from U by linear independence of the τ's. So $U \to R(U)$ is a map.

Thus for each U, we have a real 3×3 matrix $R(U)$.

For $V \in SU(2)$,

$$VU\tau_i U^\dagger V^\dagger = (VU)\tau_i(VU)^\dagger$$

or

$$V\{R_{ki}(U)\tau_k\}V^\dagger = R_{ji}(VU)\tau_j$$

or

$$R_{ki}(U)R_{jk}(V)\tau_j = R_{ji}(VU)\tau_j$$

or

$$R(V)R(U) = R(VU).$$

2) Using the identity $\tau_i\,\tau_j = \delta_{ij}\,1 + i\epsilon_{ijk}\,\tau_k$, we have

$$\begin{aligned}
U\tau_i\tau_jU^\dagger &= U\tau_iU^\dagger U\tau_jU^\dagger \\
&= R_{li}(U)\tau_l R_{mj}(U)\tau_m \\
&= R_{li}R_{mj}\{\delta_{lm} + i\epsilon_{lmn}\tau_n\} \\
&= (R^TR)_{ij}1 + i\{R_{li}R_{mj}\epsilon_{lmn}\tau_n\}.
\end{aligned}$$

Also

$$U\tau_i\tau_jU^\dagger = \delta_{ij}1 + i\epsilon_{ijk}R_{nk}\tau_n.$$
$$\Rightarrow R^TR = 1 \;\; or \;\; R \in O(3)$$

and

$$R_{li}R_{mj}\epsilon_{lmn} = \epsilon_{ijk}R_{nk}.$$

Now multiply by R_{ns} and use $R^TR = 1$. Then

$$R_{li}R_{mj}R_{ns}\epsilon_{lmn} = \epsilon_{ijs},$$

i.e. $\det R = 1$.

So

$$R \in SO(3).$$

LECTURE 3

i) $SU(2)$ and $SO(3)$ (continued)

We have shown that the image of $SU(2)$ under R is contained in $SO(3)$. We want to show the equality, i.e. $\{R(U)\} = SO(3)$.

3) To show that the map R is onto $SO(3)$:

Let $S \in SO(3)$. We know that $S = S_3(\gamma)S_2(\beta)S_3(\alpha)$ where $S_i(\phi) \equiv$ rotation by ϕ around the ith axis. We show that $U_3(\alpha)$ exists such that $R[U_3(\alpha)] = S_3(\alpha)$. Similarly one finds $U_2(\beta)$, $U_3(\gamma)$. Finally because R is a homomorphism, it follows that $U_3(\gamma)U_2(\beta)U_3(\alpha) \xrightarrow{R} S$.

Now

$$S_3(\alpha) = \begin{bmatrix} \cos\alpha & \sin\alpha & 0 \\ -\sin\alpha & \cos\alpha & 0 \\ 0 & 0 & 1 \end{bmatrix}.$$

To find $U_3(\alpha)$ such that $U_3(\alpha)\, \tau_i\, U_3(\alpha)^\dagger = S_3(\alpha)_{ji}\, \tau_j$, we first write out each term in this equation:

$$U_3(\alpha)\, \tau_1\, U_3(\alpha)^\dagger = \cos\alpha\, \tau_1 - \sin\alpha\, \tau_2,$$
$$U_3(\alpha)\, \tau_2\, U_3(\alpha)^\dagger = \sin\alpha\, \tau_1 + \cos\alpha\, \tau_2,$$
$$U_3(\alpha)\, \tau_2\, U_3(\alpha)^\dagger = \tau_3.$$

Try

$$U_3(\alpha) = e^{i\frac{\tau_3}{2}\alpha}$$
$$= \cos\frac{\alpha}{2} + i\,\tau_3\,\sin\frac{\alpha}{2}$$
$$= \begin{bmatrix} e^{\frac{i\alpha}{2}} & 0 \\ 0 & e^{-\frac{i\alpha}{2}} \end{bmatrix} \in SU(2).$$

11

$$U_3(\alpha)^\dagger = e^{-i\frac{\tau_3}{2}\alpha}$$

$$= \cos\frac{\alpha}{2} - i\,\tau_3\,\sin\frac{\alpha}{2}.$$

Then

$$U_3(\alpha)\,\tau_2\,U_3(\alpha)^\dagger = \left[\tau_1\,\cos\frac{\alpha}{2} - \tau_2\,\sin\frac{\alpha}{2}\right]\left[\cos\frac{\alpha}{2} - i\,\tau_3\,\sin\frac{\alpha}{2}\right]$$

$$= \tau_1\left(\cos^2\frac{\alpha}{2} - \sin^2\frac{\alpha}{2}\right) + \tau_2\left(-2\cos\frac{\alpha}{2}\sin\frac{\alpha}{2}\right)$$

$$= \cos\alpha\,\tau_1 - \sin\alpha\,\tau_2.$$

Similarly for the rest of the equations.

Finally we show that the map R is two-to-one.

4) To show that the map is $2 \to 1$:

(a) $R(U) = R(-U)$ so that $\pm U$ are mapped to the same $R(U)$.

(b) If $U\,\tau_i\,U^\dagger = V\,\tau_i\,V^\dagger$, then $U = \pm V$. For then,

$$V^\dagger U\,\tau_i = \tau_i\,V^\dagger U$$

or

$$[V^\dagger U,\,\tau_i] = 0.$$

But $[V^\dagger U,\,1] = 0$, so $[V^\dagger U,\,M] = 0$ for any M. Consequently

$$V^\dagger U = \alpha_0\,1.$$

Taking the determinant we have

$$\alpha_0^2 = 1$$

or

$$\alpha_0 = \pm 1.$$

This completes the proof.

ii) Subgroups

A subgroup H of a group G is a (non-empty) subset of G which itself forms a group with respect to the group composition defined on G.

Thus

$$h \in H \Rightarrow h^{-1} \in H.$$
$$h_1, h_2 \in H \Rightarrow h_1 h_2 \in H.$$
$$e \in H.$$

Examples:

1) S_3 has three distinct S_2 subgroups consisting of e and permutation of i and j alone. All are isomorphic.

2) $SO(3)$ consists of an infinite number of $SO(2)$ subgroups. A typical one will be all rotations about a fixed axis $\hat{n}$:

$$\{e^{i\hat{n} \cdot \vec{J}\theta} | \hat{n} \text{ fixed, all } \theta\},$$

$$\vec{J} \equiv \text{spin one angular momentum operator.}$$

They are all isomorphic. The isomorphism map is:

$$e^{i\hat{n} \cdot \vec{J}\theta} \rightarrow e^{i\hat{m} \cdot \vec{J}\theta}.$$

3) Real orthogonal matrices are also unitary. So $O(n) \subset U(n)$, $SO(n) \subset SU(n)$. Also, $U(n) \supset U(n-1) \supset U(n-2) \supset \cdots \supset U(1)$.

4) $G = $ the Poincaré group. The translations $\mathcal{T}_4 = \{(a, 1)\}_{a \in \mathbb{R}^4}$ form a subgroup of G. The Lorentz transformations $O(3, 1) = \{(0, \Lambda)\}$ form a subgroup of G.

iii) Cosets and invariant subgroups

Let H be a subgroup of a group G. Then the *left coset* of H with respect to an element $g \in G$ is the set $gH \equiv \{gh | h \in H\}$. (Similarly right cosets are given by $Hg = \{hg \mid h \in H\}$. The space of left (right) cosets is $\{gH\}_{g \in G} (\{Hg\}_{g \in G})$.)

We will prove the following:

(1) $G = \cup_g gH$.

(2) $gH \cap g'H$ is either $gH = g'H$ or $\emptyset$ (the null set).

(3) If $\bar{g} \in gH$, then $\bar{g}H = gH$.

Thus if we say that g_1 is equivalent to g_2, $g_1 \sim g_2$, when g_1 and g_2 are in the same left coset, then the symbol $\sim$ is an equivalence relation. That is, it is symmetric ($g_1 \sim g_2 \Rightarrow g_2 \sim g_1$), reflexive ($g \sim g$) and transitive ($g_1 \sim g_2$, $g_2 \sim g_3 \Rightarrow g_1 \sim g_3$). Also since left cosets are identical or totally disjoint, we can label them by picking one element $\bar{g}$ from each left coset gH. Then automatically, $\bar{g}H = gH$.

Similar statements hold for right cosets.

(1) is due to the fact that $e \in H$ and hence $g \in gH$.

(2) and (3) are proved by the following lemmas.

Lemma 1

g_1 and g_2 are in the same left coset iff $g_1{}^{-1}g_2 \in H$.

If: Let $g_1{}^{-1}g_2 = h \in H$.   Then $g_2 = g_1\hat{h}$ or $g_2H = g_1hH = g_1H$. ($hH = H$ because H is a group.)

Only if: Suppose $g_1, g_2 \in gH$. Then $\exists\, h_1, h_2 \in H$ such that $g_1 = gh_1$ and $g_2 = gh_2$. Consequently

$$g_1{}^{-1}g_2 = h_1{}^{-1}h_2 \in H.$$

Lemma 2

For any two cosets g_1H, g_2H, either $g_1H = g_2H$ or $g_1H \cap g_2H = \emptyset$.

Proof:

Suppose g_1H and g_2H have an element g in common.   Then $\exists\, h_1$, $h_2 \in H$ such that $g = g_1h_1 = g_2h_2$. This implies $g_1{}^{-1}g_2 = h_1h_2{}^{-1} \in H$ and (by Lemma 1) $g_1H = g_2H$.

Lemma 3

If $\bar{g} \in gH$, then $\bar{g}H = gH$.

Lemma 2 gives the result since both the cosets contain $\bar{g}$, e being in H.

A subgroup $H \subset G$ is an *invariant* or a *normal* subgroup if $gHg^{-1} = H$ for all $g \in G$.

Note that now each $g \in G$ induces an automorphism f_g of H where $f_g(h) = ghg^{-1}$.

LECTURE 4

i) Cosets and invariant subgroups (continued)

Examples:

1) Let V_n be an n-dimensional vector space. It is a group under vector addition. If V_m is an m-dimensional subspace of V_n, it is an invariant subgroup.

2) $G =$ the Poincaré group. It contains 4-dimensional translations $T_4 = \{(a,\ 1)\}$ as an invariant subgroup.

3) $G = SO(3)$, $H =$ rotations around 3rd axis $= \{R_3(\gamma)\}$. This H is not an invariant subgroup of G.

Lemma

If H is an invariant subgroup of G, the left and right cosets are the same.

So in this case one can talk of cosets without specifying left or right and denote the space of cosets by

$$G/H = \{gH\} = \{Hg\}.$$

Theorem

If H is an invariant subgroup of G, G/H is a group, the factor group of G with respect to H. The multiplication rule is set multiplication of cosets. The identity is H.

For

$$gHg'H = gg'HH = gg'H = \text{another coset};$$

$$gHH = gH \text{ or } H \text{ is the identity};$$

$$(gH)^{-1} = g^{-1}H \text{ is the inverse since } g^{-1}HgH = H.$$

Examples:

1) $G = O(3)$; $H = SO(3)$. Because $\det(ghg^{-1}) = \det h = 1$, H is an invariant subgroup of G.

If $g \notin H$, $g = Ph$ where $h \in H$ and

$$P = \begin{bmatrix} -1 & & \\ & -1 & \\ & & -1 \end{bmatrix}.$$

So $gH = PH$ and the cosets are H and PH. The group multiplication table becomes:

	H	PH
H	H	PH
PH	PH	H

This group is isomorphic to S_2.

2) $G = $ the Poincaré group, $H = T_4$. Since

$$T_4(a, \Lambda) = T_4(a, 1)(0, \Lambda) = T_4(0, \Lambda),$$

these cosets are in one-to-one correspondence with the elements of the Lorentz group $\{\Lambda\}$. Since

$$T_4(0, \Lambda)T_4(0, \Lambda') = T_4(0, \Lambda\Lambda'),$$

G/H is isomorphic to the Lorentz group.

3) $G = S_n$, $H = A_n = $ the alternating subgroup $= $ the set of all even permutations. [Any permutation s can be written as $t_1 \ldots t_k$ where t_i is a transposition. If k is even (odd), then s is even (odd)(see Lecture 11).]

Clearly A_n is a group. Also since $a \in A_n$ implies sas^{-1} is even, A_n is an invariant subgroup.

If s is odd, then $s = t_0 a$ where t_0 is any transposition and a is even. Also

$$a \text{ even} \Rightarrow aA_n = A_n$$

while

$$s \text{ odd} \Rightarrow A_n = t_0 A_n.$$

So

$$S_n/A_n = S_2.$$

ii) Conjugate elements and classes

We say g_1 and g_2 are conjugate to each other if $\exists g_0 \in G$ such that

$$g_2 = g_0 g_1 g_0^{-1}.$$

Let us write $g_1 \sim g_2$ if such a g_0 exists in G. The relation $\sim$ is symmetric, reflexive, and transitive, i.e. a) $g_1 \sim g_2$ iff $g_2 \sim g_1$; b) $g \sim g$; c) $g_1 \sim g_2$, $g_2 \sim g_3 \Rightarrow g_1 \sim g_3$. So $\sim$ is an equivalence relation. The set of all such equivalent g's forms a class.

Examples:

1. The class containing $e \in G$ is $\{e\}$.
2. $G = SO(3)$. Consider a rotation by θ around the axis $\hat{n}$:

$$g = e^{i\hat{n}\cdot\vec{J}\theta} = g(\hat{n}, \theta).$$

If R is a rotation, then

$$RJ_i R^{-1} = R_{ji} J_j.$$

So

$$RgR^{-1} = \exp\left\{i\, \hat{n}_i RJ_i R^{-1}\theta\right\}$$
$$= \exp\left\{i\, R_{ji}\hat{n}_i J_j\theta\right\}.$$

So the class containing this g is that of rotations about every axis by the same angle θ.

iii) Simple groups

Definition. A group G is *simple* if it has no invariant subgroup besides the identity and itself.

Examples:

1. S_2 is simple.
2. $SO(3)$ and the group of proper orthochronous Lorentz transformations. $\mathcal{L}_+^\uparrow$ are simple. [We omit the proofs.]
3. $SU(2)$ is not simple as it has the invariant subgroup

$$Z_2 = \left\{\begin{bmatrix} 1 & 0 \\ 0 & 1 \end{bmatrix}, -\begin{bmatrix} 1 & 0 \\ 0 & 1 \end{bmatrix}\right\}.$$

4. $SU(n)$ is not simple as it has the invariant subgroup

$$Z_n = \left\{ \exp\left(i\frac{2\pi}{n}k \right) 1 \mid k = 0,\ 1,\ \ldots,\ n-1 \right\}.$$

5. The group $SO(2)$ is Abelian, so every discrete subgroup of $SO(2)$ is invariant. Therefore, $SO(2)$ is not simple. [See next lecture for the definition of Abelian groups.]

6. The group $SL(2,\mathbb{C})$ is $\{g\}$ where the element g is a 2×2 complex matrix with det $g = +1$. It is not simple because the subgroup

$$Z_2 = \left\{ \begin{bmatrix} 1 & 0 \\ 0 & 1 \end{bmatrix}, - \begin{bmatrix} 1 & 0 \\ 0 & 1 \end{bmatrix} \right\}$$

is invariant.

LECTURE 5

i) Abelian and semi-simple groups

Definition. A group $G = \{g_\alpha\}$ is Abelian if $g_\alpha g_\beta = g_\beta g_\alpha$, $\forall\, g_\alpha,\, g_\beta$.

Examples:

1) Translations in $\mathbb{R}^n$ are Abelian.
2) $SO(2)$ is Abelian.

Definition. A group G is *semi-simple* if it has no Abelian invariant subgroup besides the identity and itself.

Examples:

1) Any discrete subgroup of $SO(2)$ is Abelian and invariant. So $SO(2)$ is not semi-simple.
2) $SU(n)$ is not semi-simple since

$$Z_n = \left\{ e^{2\pi i k/n}\, 1 \mid k = 0,\, 1,\, \ldots,\, n-1 \right\}$$

is an Abelian invariant subgroup.
3) The Poincaré group is not semi-simple since translations form an Abelian invariant subgroup.

ii) Representations of a group

Given a group $G = \{g\}$, a representation $\Gamma = \{T(g)\}$ of G is a set of linear operators on a vector space which forms a group under the usual product rule of linear operators such that

$$g \xrightarrow{\ T\ } T(g)$$

is a homomorphism. The vector space V on which Γ acts is the carrier or support space of the representation.

We will freely interchange the use of linear operators with matrices in view of their well-known correspondence.

The representation is called *faithful* if T is an isomorphism.

Examples:

1) $SO(3)$ is a representation of $SU(2)$ which is not faithful. However, $SU(2)$ is not a representation of $SO(3)$. The spin j representation of $SU(2)$ is $\Gamma^{(j)} = \{D^{(j)}\}$ where $D^{(j)}$ are rotation matrices for angular momentum $j = 0, 1/2, 1, \ldots$. For j half an odd integer, the representation is faithful for $SU(2)$, but not so for j an integer. For j an integer, they are representations of $SO(3)$ and when $j \neq 0$, they are faithful representations of $SO(3)$. [For proof, see Lectures 3 and 25.]

2)

$$P = \begin{pmatrix} 1\ 2\ 3\ 4 \\ 2\ 3\ 4\ 1 \end{pmatrix},$$

P^2, P^3, $P^4 = e$ form a group. Now the fourth roots of unity are $e^{2\pi i k/4}$, $k = 0,\ 1,\ 2,\ 3$. Define

$$T(P^k) = e^{2\pi i k/4} 1.$$

$T(P^k)$ is a linear operator on the one-dimensional complex vector space $\mathbb{C}^1 = \{z\}$. $\Gamma = \{T(P^k)\}$ is a faithful representation of the above group.

3) A four-dimensional representation of $SO(3)$ is given by

$$SO(3) \ni R \to \begin{bmatrix} R & 0 \\ 0 & 1 \end{bmatrix}.$$

A $3n$-dimensional representation is given by

$$R \to \begin{bmatrix} R & 0 & . & . & 0 \\ 0 & R & . & . & . \\ . & . & R & . & . \\ . & . & . & . & . \\ 0 & . & . & . & R \end{bmatrix}.$$

4) Suppose S is a fixed 3×3 non-singular matrix and $R \in SO(3)$. Then the map $R \to SRS^{-1}$ gives a representation $\{SRS^{-1}\}$ of $SO(3)$.

iii) The regular representations

Consider the set V of all functions from G to $\mathbb{C}$:

$$G \ni g \xrightarrow{\ f \in V\ } f(g) \in \mathbb{C}.$$

Define for α, β complex numbers and f_1, $f_2 \in V$ the function $\alpha f_1 + \beta f_2$ by

$$(\alpha f_1 + \beta f_2)(g) = \alpha f_1(g) + \beta f_2(g).$$

This makes V into a complex vector space.

For each $g \in G$, define the linear operator $T(g)$ on V as follows: For any f, $T(g)f$ is the function defined by

$$[T(g)f](g') = f(g^{-1}g').$$

For $g, \bar{g} \in G$,

$$\begin{aligned}
[T(g)T(\bar{g})f](g') &= [T(\bar{g})f](g^{-1}g') \\
&= f(\bar{g}_{-1}g^{-1}g') \\
&= f[(g\bar{g})^{-1}g'] \\
&= [T(g\bar{g})f](g').
\end{aligned}$$

So

$$T(g)T(\bar{g}) = T(g\bar{g}),$$

and $\{T(g)\}$ is a representation of G, called the *left regular representation*.

Now define $S(g)$ on V by

$$[S(g)f](g') = f(g'g).$$

Then

$$\begin{aligned}
[S(g)S(\bar{g})f](g') &= [S(\bar{g})f](g'g) \\
&= f(g'g\bar{g}) \\
&= [S(g\bar{g})f](g')
\end{aligned}$$

or

$$S(g)S(\bar{g}) = S(g\bar{g}).$$

Thus $\{S(g)\}$ is a representation of G, called the *right regular representation*.

For finite groups, define the scalar product

$$(\alpha, \ \beta) = \sum_g \alpha^*(g)\beta(g)$$

for $\alpha, \ \beta \in V$.

Then

$$(T(\bar{g})\alpha, T(\bar{g})\beta) = \Sigma_g \alpha^*(\bar{g}^{-1}g)\beta(\bar{g}^{-1}g)$$
$$= (\alpha, \beta).$$

Thus $T(g)$'s (and similarly $S(g)$'s) are unitary in this scalar product.

Definition. Given two representations

$$\Gamma_1 = \{D^{(1)}(g)\} \text{ and } \Gamma_2 = \{D^{(2)}(g)\}$$

on vector spaces V_1, V_2, they are *equivalent* if there exists a non-singular $(1-1, \text{onto})$ linear operator S from V_1 onto V_2 such that

$$SD^{(1)}(g)S^{-1} = D^{(2)}(g).$$

LECTURE 6

i) Reducibility of representations

Suppose one is given a representation Γ on a vector space V. A subspace V_0 of V is invariant under a linear transformation T if $TV_0 \subset V_0$, that is, if $x_0 \in V_0$ implies $Tx_0 \in V_0$. A subspace V_0 is invariant under $\Gamma = \{D(g)\}$ if it is invariant under every $D(g)$. A representation Γ is reducible if there is a subspace $V_0 (\neq \{0\}$ or $V)$ of V which is invariant under Γ.

Suppose Γ is reducible. Choose a basis $e_1, \ldots, e_n$ for V where $e_1, \ldots, e_m$ span V_0. The matrix $D(g)$ in this basis is of the form

$$D(g) = \begin{bmatrix} D_1(g) & \alpha(g) \\ 0 & D_2(g) \end{bmatrix}$$

where $D_1(g)$ is $m \times m$, $\alpha(g)$ is $m \times (n-m)$ and $D_2(g)$ is $(n-m) \times (n-m)$. Conversely if $D(g)$ does take this form in some basis, then vectors of the form $(\xi_1, \ldots, \xi_m, 0, \ldots, 0)$ are clearly invariant under Γ, that is, they span an invariant subspace. Thus Γ is reducible iff all $D(g)$'s are of the above form in some basis.

Note that

$$D(g_1)D(g_2) = \begin{bmatrix} D_1(g_1) & \alpha(g_1) \\ 0 & D_2(g_1) \end{bmatrix} \begin{bmatrix} D_1(g_2) & \alpha(g_2) \\ 0 & D_2(g_2) \end{bmatrix}$$

$$= \begin{bmatrix} D_1(g_1)D_1(g_2) & D_1(g_1)\alpha(g_2) + \alpha(g_1)D_2(g_2) \\ 0 & D_2(g_1)D_2(g_2) \end{bmatrix}$$

$$= D(g_1 g_2)$$

by the group property of $D(g)$'s. Therefore

$$D_1(g_1)D_1(g_2) = D_1(g_1 g_2)$$
$$D_2(g_1)D_2(g_2) = D_2(g_1 g_2).$$

23

Thus $\Gamma_1 = D_1(g)$ and $\Gamma_2 = D_2(g)$ are themselves representations.

Examples:

1) Let $\mathcal{T}_1 = \{T(a)\}_{a \in \mathbb{R}^1}$ be translations on the vector space $V_1 = \{x\}_{x \in \mathbb{R}^1}$ such that

$$T(a)x = x + a.$$

The $T(a)$'s are not linear operators on V_1. Consider now

$$\Gamma = \left\{ D(a) = \begin{bmatrix} 1 & a \\ 0 & 1 \end{bmatrix} \right\}.$$

Then

$$D(a)D(b) = \begin{bmatrix} 1 & a \\ 0 & 1 \end{bmatrix} \begin{bmatrix} 1 & b \\ 0 & 1 \end{bmatrix} = \begin{bmatrix} 1 & a+b \\ 0 & 1 \end{bmatrix}$$
$$= D(a+b).$$

Thus Γ is a reducible representation of $\mathcal{T}_1$.

The two-dimensional translation group is $\mathcal{T}_2 = \{T(a_1, a_2)\}$ where

$$T(a_1, a_2) \begin{pmatrix} x_1 \\ x_2 \end{pmatrix} = \begin{pmatrix} x_1 + a_1 \\ x_2 + a_2 \end{pmatrix}.$$

A two-dimensional representation for $\mathcal{T}_2$ is

$$\left\{ D(a_1, a_2) = \begin{bmatrix} 1 & a_1 + ia_2 \\ 0 & 1 \end{bmatrix} \right\}.$$

A four-dimensional representation is

$$\left\{ \begin{bmatrix} 1 & a_1 \\ 0 & 1 \end{bmatrix} \otimes \begin{bmatrix} 1 & a_2 \\ 0 & 1 \end{bmatrix} \right\} = \begin{bmatrix} 1 & a_2 & a_1 & a_1 a_2 \\ 0 & 1 & 0 & a_1 \\ 0 & 0 & 1 & a_2 \\ 0 & 0 & 0 & 1 \end{bmatrix}.$$

2) Suppose one is given two representations $\Gamma_i = \{D_i(g)\}$ on vector spaces V_i where D_i are matrices. The direct sum $\Gamma_1 \oplus \Gamma_2$ is the reducible representation given by

$$\left\{ \begin{bmatrix} D_1(g) & 0 \\ 0 & D_2(g) \end{bmatrix} \right\}.$$

The preceding construction can be stated in a coordinate-free way as follows: If one is given two vector spaces V_1 and V_2, their direct sum $V =$

$V_1 \oplus V_2$, is the linear combination of vectors in V_1 and V_2 in a well-known sense.

The dimension of V, dim V, is dim V_1+dim V_2. If T_1 is a linear operator on V_1, and T_2 is a linear operator on V_2, then the direct sum $T = T_1 \oplus T_2$ is a linear operator on V defined as follows: If $z \in V$, then $z = x + y$, where $x \in V_1$ and $y \in V_2$. We set

$$Tz = T_1 x + T_2 y.$$

Now given representations $\Gamma_i = \{T_i(g)\}$ on V_i, their direct sum Γ is $\Gamma_1 \oplus \Gamma_2 = \{T(g) = T_1(g) \oplus T_2(g)\}$. Note that

$$
\begin{aligned}
T(g_1)T(g_2)z &= T(g_1)(T_1(g_2)x + T_2(g_2)y) \\
&= T_1(g_1)T_1(g_2)x + T_2(g_1)T_2(g_2)y \\
&= T_1(g_1 g_2)x \qquad + T_2(g_1 g_2)y \\
&= T(g_1 g_2)z.
\end{aligned}
$$

Here $z \in V_1 \oplus V_2$, $x \in V_1$, $y \in V_2$. Thus Γ is also a representation.

Now choose a basis $e_1, \ldots, e_m, f_1, \ldots, f_n$ where e_i span V_1, f_i span V_2. Then in this basis,

$$
T(g) = \begin{bmatrix} D_1(g) & 0 \\ 0 & D_2(g) \end{bmatrix},
$$

or Γ is reducible.

ii) Full or complete reducibility

Let $\Gamma = D(g)$ be a reducible representation on a vector space V with invariant subspace V_1. Now if there exists another invariant subspace V_2 such that $V = V_1 \oplus V_2$, then Γ is *fully reducible* into a direct sum $\Gamma_1 \oplus \Gamma_2$ where Γ_i is the *restriction* of Γ to V_i. [In the preceding discussion, $\Gamma_i = \{D_i(g)\}$.]

Now Γ_1 may or may not be reducible on V_1. If it is reducible, we can try writing V_1 itself as the direct sum of two invariant subspaces.

iii) Irreducible representations

Definition. A representation Γ on a vector space V is irreducible (IRR) if V has no invariant subspace besides $\{0\}$ and V itself.

Examples:

1) The representation

$$\left\{ \begin{bmatrix} 1 & a \\ 0 & 1 \end{bmatrix} \right\}$$

of T_1 is reducible, but not fully reducible. Here

$$V_1 = \left\{ \begin{pmatrix} x \\ 0 \end{pmatrix} \right\}, \; x \in \mathbb{R}^1$$

is invariant. The general form of V_2 (if it exists) will be

$$V_2 = \left\{ \begin{pmatrix} \lambda y \\ y \end{pmatrix} \mid \lambda \text{ fixed}, y \in \mathbb{R}^1 \right\}.$$

But

$$\begin{bmatrix} 1 & a \\ 0 & 1 \end{bmatrix} \begin{pmatrix} \lambda y \\ y \end{pmatrix} = \begin{pmatrix} y(\lambda + a) \\ y \end{pmatrix} \notin V_2 \text{ if } a \neq 0.$$

So V_2 is not invariant under T_1.

Theorem

2) Let H be a Hilbert space with scalar product: $(\cdot, \cdot)$. Let $\Gamma = \{U(g)\}$ be a unitary representation of G on H:

$$(x, \; U(g)y) = (U^\dagger(g)x, \; y) = (U^{-1}(g)x, \; y).$$

Then Γ is the direct sum of irreducible representations: $\Gamma = \oplus_i \Gamma_i$. Also, if $H = \oplus_i H_i$ is the corresponding decomposition of H, then this direct sum of H can be chosen to be an orthogonal sum, i.e., if $x_\alpha \in H_\alpha$, $x_\beta \in H_\beta$, then $(x_\alpha, \; x_\beta) = 0$ if $\alpha \neq \beta$.

Proof :

Note that:

a) If H_0 is a subspace of H, then we have the decomposition $H = H_0 \oplus H_1$ where the sum is an orthogonal direct sum. H_1 is the orthogonal complement of H_0 in H.

b) If U is unitary, defined on H, and H_0 is invariant under U, then so is H_1.

For since H_0 is invariant under U and U is unitary, $\exists \, e_1, \ldots, e_m$ spanning H_0 such that $Ue_i = \lambda_i e_i$, $|\lambda_i| = 1$. This implies that $U^{-1}e_i = \lambda_i^{-1} e_i$

so that H_0 is invariant under U^{-1}. Now let $x_0 \in H_0$, $x_1 \in H_1$. By definition, $(x_1, x_0) = 0$. To prove $(Ux_1, x_0) = 0$, we note that $(Ux_1, x_0) = (x_1, U^\dagger x_0) = (x_1, U^{-1} x_0) = (x_1, x_0') = 0$ since $x_0' \in H_0$.

c) Thus if H_0 is invariant under Γ, we can write $H = H_0 \oplus H_1$ where H_1 is orthogonal to H_0 and invariant under Γ. If $H_i(i = 0, 1)$ has an invariant subspace under Γ, one repeats the process.

LECTURE 7

i) Schur's lemma

Given two vector spaces V and W, we can think of linear operators L from V to W:

$$V \ni x \xrightarrow{L} Lx \in W$$

such that

$$L(x_1 + x_2) = Lx_1 + Lx_2$$

and

$$L(\lambda x) = \lambda Lx, \quad \lambda = \text{any complex number.}$$

For example $V = \{(\xi_1, \ldots, \xi_m)\}$, $W = \{(\eta_1, \ldots, \eta_m)\}$ and L is an $n \times m$ matrix.

Let $L : V \to W$ be a linear operator. The null space $N(L)$ of L is $\{x | Lx = 0 \in W\}$ and the range $R(L)$ of L is $\{y | y = Lx, \ x \in V\}$.

1. The map $L : V \to W$ is $1-1$ iff $N(L)$ consists of the zero vector. If: $Lx_1 = Lx_2 \Rightarrow L(x_1 - x_2) = 0 \Rightarrow x_1 - x_2 \in N(L) \Rightarrow x_1 = x_2$ since $N(L)$ consists of the zero vector. $Only \ if$: $\xi \in N(L) \Rightarrow Lx = L(x + \xi)$.

2. The map $L : V \to W$ is onto iff $R(L) = W$.

3. L is invertible iff $N(L) = 0$ and $R(L) = W$.

4. If L is invertible, $\dim W = \dim V$.

Proofs to 2-4:

Here 2 and 3 are obvious while 4 can be proved as follows.

Choose a basis $x_1, \ldots, x_m \in V$. If $y \in W$, $\exists \, x \in V$ such that $y = Lx$. Then $x = \sum_i \xi_i x_i \Rightarrow y = \sum_i \xi_i (Lx_i) \Rightarrow Lx_i$ is complete in W. Also

$$\sum_i \eta_i Lx_i = 0 \Rightarrow \eta_i = 0$$

29

because

$$L\left(\sum \eta_i x_i\right) = 0 \Rightarrow \sum_i \eta_i x_i \in N(L) \Rightarrow \sum_i \eta_i x_i = 0 \Rightarrow \eta_i = 0$$

by linear independence of $\{x_i\}$. Thus the $\{Lx_i\}$ are linearly independent. As $\{Lx_i\}$ is complete as well, $\{Lx_i\}$ is a basis for W. Hence dim $V =$ dim W.

Statement of Schur's Lemma:

1) Let $\Gamma_A = \{A(g)\}$ and $\Gamma_B = \{B(g)\}$ be two IRR's of a group $G = \{g\}$ on complex vector spaces V_A, V_B. Let $L : V_A \to V_B$ be a linear operator such that

$$LA(g) = B(g)L, \quad g \in G.$$

(One says that L intertwines the two representations.) Then either $L = 0$ or L is invertible so that

$$B(g) = LA(g)L^{-1}$$

and the two representations are equivalent. (The possibility that L is singular, but not zero is excluded.)

2) Let $\Gamma = \{D(g)\}$ be an IRR on a vector space V, and let $L : V \to V$ be a linear operator such that $[L, D(g)] = 0$. Then L is a multiple of the unit operator.

See H. Weyl: *Classical Groups*, Chap. 5. The following proof is from this chapter.

1) We can assume that $L \neq 0$. We then show that $N(L)$ and $R(L)$ are invariant subspaces of Γ_A and Γ_B. By irreducibility it will follow that $N(L) = \{0\}$ or V_A and $R(L) = \{0\}$ or V_B. Then $L \neq 0 \Rightarrow N(L) = \{0\}$ and $R(L) = V_B \Rightarrow L$ is invertible.

Let $x \in N(L)$, then

$$LA(g)x = B(g)Lx = 0$$

or

$$A(g)x \in N(L).$$

But this implies that $N(L)$ is invariant under Γ_A.

Let $y \in R(L)$, then $\exists \, x \in V_A$ such that $y = Lx$. Therefore

$$B(g)y = B(g)Lx$$
$$= L(A(g)x).$$

Consequently, $B(g)y \in R(L)$, and this implies that $R(L)$ is invariant under Γ_B.

2) L has at least one eigenvector x with $Lx = \lambda x$. Therefore $(L - \lambda 1)x = 0 \Rightarrow (L - \lambda 1)$ is singular. But $[L - \lambda 1, D(g)] = 0$, so by 1), $L - \lambda 1 = 0$ (since $L - \lambda 1$ is singular) or $L = \lambda 1$.

Example :

All IRR's of an Abelian group are one-dimensional. For let $\Gamma = D(a)$ be such an IRR. Then $D(a)D(b) = D(b)D(a) \Rightarrow$ for any $a_0, D(a_0)$ commutes with every $D(a)$. So 2) $\Rightarrow D(a_0) = \lambda(a_0)1$. This is true for all a_0. Since Γ is IRR, here 1 must be a 1×1 matrix.

If $T_n = \{a\}_{a \in \mathbb{R}^n}$ is the n-dimensional translation group, its IRR's are given by

$$a = (a_1, \ldots, a_n) \rightarrow \exp \sum \lambda_i a_i, \ (\lambda_1, \cdots, \lambda_n) \in \mathbb{C}^n$$

with λ_i fixed in a given IRR. [Prove this as an exercise. Here $\mathbb{C}^n = n$ - dimensional complex vector space.]

ii) Operations with representations and with groups

1) Let $\Gamma = D(g)$ be a representation of a group G in terms of matrices.

$$(a) \ D(g_1)^* D(g_2)^* = [D(g_1)D(g_2)]^*$$
$$= D(g_1 g_2)^*$$
$$\Rightarrow g \rightarrow D(g)^*$$

gives a representation $\Gamma^* = \{D(a)^*\}$, the complex conjugate of Γ.

$$(b) \ D(g_1)^T D(g_2)^T = [D(g_2)D(g_1)]^T$$
$$= D(g_2 g_1)^T.$$

So

$$D(g_1^{-1})^T D(g_2^{-1})^T = D((g_1 g_2)^{-1})^T$$
$$\Rightarrow g \rightarrow D(g^{-1})^T$$

gives a representation called the representation contragradient to Γ.

$$(c) \ g \rightarrow D(g^{-1})^\dagger$$

also gives a representation, which can be called $\Gamma^\dagger$. If the representation is unitary,

$$D(g^{-1})^\dagger = D(g^{-1})^{-1} = D(g).$$

So (c) is the same as Γ, and (b) is the same as (a).

Γ and Γ^* may or may not be equivalent.

Example:

For $SU(2)$ or $SO(3)$, Γ and Γ^* are equivalent for any Γ. [See Problem 19.]

For $SU(3)$, $G = \{g\}$ where g is a 3×3 matrix with $\det g = 1$ and $g^\dagger = g^{-1}$. The representation $\underline{3}$ for $SU(3)$ associates the matrix g itself with g, i.e.,

$$SU(3) \ni g \to D(g) \equiv g \in \underline{3}.$$

So

$$SU(3) \ni g \to D(g)^* \equiv g^* \in \underline{3}^*.$$

Now

$$z = \begin{bmatrix} e^{2\pi i/3} & & \\ & e^{2\pi i/3} & \\ & & e^{2\pi i/3} \end{bmatrix} \in SU(3).$$

We have $D(z) = z$, but

$$D(z)^* = z^* = \begin{bmatrix} e^{4\pi i/3} & & \\ & e^{4\pi i/3} & \\ & & e^{4\pi i/3} \end{bmatrix}.$$

Since $D(z)$, $D(z)^*$ have different spectra (their eigenvalues differ), they cannot be related by a similarity transformation. This implies that $\underline{3}$ and $\underline{3}^*$ are not equivalent. Therefore, $*$ is an outer automorphism. (An inner automorphism would be a map via conjugation by an element of G itself.)

2) Similarity transformations on representations give equivalent representations.

3) Direct sums of representations give representations.

iii) Direct product of representations

Consider two vector spaces V_i $(i = 1, 2)$, V_1 with basis $e_1, \ldots, e_m$ and V_2 with basis $f_1, \ldots, f_n$. Then $V_1 \otimes V_2$ has a basis

$$e_i \otimes f_j, \ i = 1, 2, \ldots, m; \ j = 1, 2, \ldots, n.$$

A general element in $V_1 \otimes V_2$ is

$$x = \sum a_{ij} \, e_i \otimes f_j.$$

The dimension dim $[V_1 \otimes V_2]$ is mn.

If L_i are linear operators on V_i, then we define $L = L_1 \otimes L_2$ by

$$Lx = \sum_{i,j} a_{ij} (L_1 e_i) \otimes (L_2 f_j).$$

If $M = M_1 \otimes M_2$ is another linear operator, then clearly

$$ML = (M_1 L_1) \otimes (M_2 L_2).$$

Suppose $D^{(i)}$ is the matrix of L_i in the above basis:

$$L_1 e_i = D_{ji}^{(1)} e_j,$$
$$L_2 f_i = D_{ji}^{(2)} f_j.$$

Then the matrix of L in basis $\{e_i \otimes f_j\}$ is defined by

$$L(e_i \otimes f_j) = D_{i'j',ij} e_{i'} \otimes f_{j'}$$
$$= L_1 e_i \otimes L_2 f_j$$
$$= D_{i'i}^{(1)} D_{j'j}^{(2)} e_{i'} \otimes f_{j'}.$$

Comparing the two expressions gives

$$D_{i'j',ij} = D_{i'i}^{(1)} D_{j'j}^{(2)}.$$

If we enumerate the basis in the order $e_1 \otimes f_1, e_1 \otimes f_2, \ldots, e_1 \otimes f_n; e_2 \otimes f_1,$ $\ldots, e_2 \otimes f_n; \ldots, e_m \otimes f_n$, then

$$D = \begin{bmatrix} D_{11}^{(1)} \, D^{(2)} & D_{12}^{(1)} \, D^{(2)} & . & . & D_{1m}^{(1)} \, D^{(2)} \\ D_{21}^{(1)} \, D^{(2)} & & . & . & D_{2m}^{(1)} \, D^{(2)} \\ & & & & \\ . & . & . & & \\ . & . & . & & \\ . & . & . & & \\ D_{m1}^{(1)} \, D^{(2)} & & . & . & D_{mm}^{(1)} \, D^{(2)} \end{bmatrix}.$$

Thus D is the Kronecker product $D^{(1)} \otimes D^{(2)}$ of $D^{(1)}$ and $D^{(2)}$.

If S_i, D_i are matrices, then recall results

$$(S_1 \otimes S_2)(D_1 \otimes D_2) = S_1 D_1 \otimes S_2 D_2$$

and

$$Tr\ D \equiv \sum_{i,j} D_{ij,\ ij} = Tr\ D^{(1)}\ Tr\ D^{(2)}.$$

Given two representations $\Gamma_i = \{D^{(i)}(g)\}$, the direct product of the two representations $\Gamma = \Gamma_1 \otimes \Gamma_2$ is $\Gamma = \{D^{(1)} \otimes D^{(2)}\}$. Γ is a representation of the group (see product rule above). One can define the direct product for any number of representations in this way. Γ may not be irreducible even if the Γ_i's are.

The Clebsch-Gordan Problem

Consider representations Γ_i of G on vector spaces V_i. Then $\Gamma = \Gamma_1 \otimes \Gamma_2$ is defined on $V_1 \otimes V_2 = V$. Assume Γ can be written as the direct sum of IRR's. Then we can pose the following problems:

a) Write Γ as the direct sum $\Gamma = \oplus \Gamma_i$ where Γ_i are IRR.

b) If Γ_α are acting on $V^\alpha \subset V$, find a basis for V^α in terms of those in V_i and V_2.

Example:

$SU(2)$:

Consider $\underline{IRR}$

$$V_{j_1} \text{ with basis } |j_1\ m_1\rangle, \ m_1 = -j_1, \ \cdots, \ +j_1 \qquad \Gamma_{j_1}$$

and

$$V_{j_2} \text{ with basis } |j_2\ m_2\rangle, \ m_2 = -j_2, \ \cdots, \ +j_2 \qquad \Gamma_{j_2}.$$

Then

$$\Gamma_{j_1} \otimes \Gamma_{j_2} = \oplus_{j=|j_1-j_2|}^{j_1+j_2} \Gamma_j.$$

A basis for Γ_j is $|j_1\ j_2\ j\ m\rangle$, $m = -j, \ \cdots, \ +j$, with

$$|j_1\ j_2\ j\ m\rangle = \sum_{m=m_1+m_2} C(j_1\ j_2\ j, m_1\ m_2\ m)|j_1\ m_1\rangle \otimes |j_2\ m_2\rangle.$$

The $C(j_1\ j_2\ j,\ m_1\ m_2\ m)$ are called Clebsch-Gordan coefficients.

iv) Direct products of groups

Let $G = \{g\}$ and $H = \{h\}$ be groups. Then the direct product group is

$$G \times H = \{(g,\ h)|\ g \in G,\ h \in H\},$$

with multiplication rule

$$(g_1,\ h_1)(g_2,\ h_2) = (g_1 g_2,\ h_1 h_2).$$

If $\{D(g)\}$, $\{S(h)\}$ are representations of G and H, then

$$\Gamma = \{D(g) \otimes S(h)\}$$

($\otimes$ is the Kronecker product) is a representation of $G \times H$. However, Γ may not be faithful.

Example:

$SU(2) \times SU(2)$: Consider the mapping of group products into Kronecker products:

$$(g,h) \rightarrow g \otimes h.$$

Then

$$(1,1) \rightarrow 1 \otimes 1 \ and$$
$$(-1,-1) \rightarrow (-1) \otimes (-1) \equiv 1 \otimes 1.$$

Consequently certain different elements in the group product can be the same in the Kronecker product.

PART 2
FINITE GROUPS

LECTURE 8

i) Definitions and simple results

A group G is finite, of order n, if it has n distinct elements. If not, it is an infinite group.

Examples:

1) S_n is of order $n!$.
2) $O(3)$ is an infinite group.

Definition. Let $X = \{a_i\}$ be a subset of the group G. Form all possible products $a_\alpha a_\beta$, ... If these products exhaust the elements of G, then X is a system of generators for G.

Definition. A group G with a system of generators consisting of one element is called a *cyclic* group.

Examples:

1) The cube roots of unity under multiplication form a group $G = \{1, z, z^2\}$. It follows that z or z^2 are generators for G.
2) S_n: Consider the cyclic permutation

$$P = \begin{pmatrix} 1 \, 2 \ldots n \\ 2 \, 3 \ldots 1 \end{pmatrix}.$$

It generates the group $(P, P^2, \ldots, P^{n-1}, P^n = e)$. It is the cyclic group of order n, a subgroup of S_n.

Theorem (Lagrange)

Let G be a finite group, of order n, and H be a subgroup of order m. Then $n = pm$ where p is an integer.

Proof :

$$G = \cup_{g \in G} gH = g_1 H \cup g_2 H \ldots \cup g_p H \qquad (*)$$

where

$$g_i H \cap g_j H = \emptyset \quad (i \neq j)$$

and p is the number of distinct (left) cosets. Comparing both sides in $(*)$, we see that the left-hand side has n elements and the right-hand side has pm elements. Consequently $n = pm$ where p is an integer.

Lemma

If G is of order n, and $g \in G$, then g^n is the identity.

Let $g \neq e$. Consider g, g^2, g^3, $\cdots$. Since G is finite, this sequence must double back on itself. This implies that $\exists$ integers ρ, σ with $\rho > \sigma$ such that $g^\rho = g^\sigma$. Then

$$g^{\rho - \sigma} = e = g^m$$

where

$$\rho - \sigma \equiv m.$$

Let m be the smallest such integer with $g^m = e$. Then g generates a subgroup of order m in G, and since the order of the group n is pm (by the previous theorem), we have

$$g^n = (g^m)^p = e^p = e.$$

Lemma

A group of prime order n is cyclic.

For $g \neq e$ (we assume $n \neq 1$), by the previous lemma, $\exists$ m such that $g^m = e$, and g generates a cyclic group of order m. This implies that m divides n by Lagrange's theorem and that n/m is an integer. As n is prime, $m = n$ or $m = 1$. But m cannot be 1 since $g \neq e$. So $m = n$.

Therefore the cyclic group $(e, g, g^2, \ldots, g^{n-1})$ has the same number of elements as the original group, hence it is the original group.

The order of the cyclic group generated by g in any group is sometimes called the period of g.

Theorem

Every finite group of order n is isomorphic to a subgroup of S_n. Let $G = \{e = A_1,\ \ldots,\ A_n\}$. Give distinct numbers $\in \{1, 2, \cdots, n\}$ to A_i, that is, construct a $1-1$ map $N : \{A_1,\ A_2,\ \ldots,\ A_n\} \to \{1,\ 2,\ \ldots,\ n\}$. Consider the map

$$G \ni B \to \begin{pmatrix} N(A_1) & \cdots & N(A_n) \\ N(BA_1) & \cdots & N(BA_n) \end{pmatrix} = P_B \in S_n.$$

This map defines an isomorphism between G and a subgroup of S_n. That the map is a homomorphism follows from

$$P_C P_B = \begin{pmatrix} N(A_1) & \cdots & N(A_n) \\ N(CBA_1) & \cdots & N(CBA_n) \end{pmatrix} = P_{CB}.$$

Further

$$P_C = P_B \Rightarrow N(CA_i) = N(BA_i) \Rightarrow C = B$$

since each element of G has a distinct number. This proves the isomorphism property as well.

The map above is not unique. It depends on how the elements of G are numbered.

LECTURE 9

i) Representation theory of finite groups

Recall that every unitary representation is completely reducible into a direct sum of IRR's. Consequently every representation equivalent to a unitary one is reducible into a direct sum of IRR's.

We use the following repeatedly below : If $G = \{g\}$ is a finite group and f is a function on G, then

$$\sum_g f(g) = \sum_g f(gs)$$
$$= \sum_g f(sg), \ \forall \ s \in G.$$

Theorem

Every representation $\Gamma = \{D(g)\}$ of a finite group $G = \{g\}$ is equivalent to a unitary representation and hence is a direct sum of IRR's.

Assume that the D's are given as matrices. Let

$$C = \frac{1}{n} \sum_g D(g)^\dagger D(g)$$

where n is the order of G. Then

1) $C^\dagger = C$.

2) All eigenvalues of C are strictly positive. For if x is any vector, the equation

$$(x, \ Cx) = \frac{1}{n} \sum_g (D(g)x, \ D(g)x) \geq 0$$

implies that no eigenvalue is negative. Also there is no zero eigenvalue. For if $Cx = 0$, then $(x, Cx) = 0 \Rightarrow D(g)x = 0 \Rightarrow x = 0$ since $D(g)$ has an

43

inverse $[= D(g^{-1})]$. Thus C has a square root, $C^{1/2}$, which is hermitian and invertible.

3) $D(s)^{\dagger}CD(s) = \frac{1}{n}\sum_g D(gs)^{\dagger}D(gs) = C = C^{1/2}C^{1/2}$.

If we write this as

$$\left[C^{-1/2}D(s)^{\dagger}C^{1/2}\right]\left[C^{1/2}D(s)C^{-1/2}\right] = 1,$$

we see that $\{D(s)\}$ is equivalent to the unitary representation

$$\{\mathfrak{D}(s) = C^{1/2}D(s)C^{-1/2}\} :$$

$$\mathfrak{D}(s)^{\dagger}\mathfrak{D}(s) = 1.$$

Orthogonality and Normalization Relations

Let $\Gamma = \{D(g)\}$ be an IRR. Then

$$P = \frac{1}{n}\sum_g D(g)AD(g^{-1}) = \lambda_A\, 1$$

where A is any matrix independent of g. Here λ_A is a number and n is the order of the group.

Proof:

$$D(s)P = \frac{1}{n}\sum_g D(sg)AD(g^{-1})$$

$$= \frac{1}{n}\sum_{g'} D(g')AD(g'^{-1})D(s)$$

$$= PD(s).$$

[In the second step we set $g' = sg$ and used the invariance of the summation.] Then by Schur's lemma,

$$P = \lambda_A\, 1.$$

Let $D(s)$ be a $\sigma \times \sigma$ matrix. A basis for $\sigma \times \sigma$ matrices is

$$\{M_{k\ell}|\ k = 1,\ \ldots,\ \sigma;\ \ell = 1,\ \ldots,\ \sigma\} \text{ where } (M_{k\ell})_{ij} = \delta_{ki}\,\delta_{\ell j}.$$

Set $A = M_{k\ell}$ and denote λ_A by $\lambda_{k\ell}$. Then

$$\lambda_{k\ell}\delta_{ij} = \frac{1}{n}\sum_g [D(g)]_{im}\,[M_{k\ell}]_{mn}\,\left[D(g^{-1})\right]_{nj}$$

$$= \frac{1}{n}\sum_g [D(g)]_{ik}\,\left[D(g^{-1})\right]_{\ell j}.$$

Taking the trace,

$$\lambda_{k\ell}\,\sigma = \frac{1}{n}\sum_g \delta_{k\ell} = \delta_{k\ell},$$

we have the following.

Theorem

If $\Gamma = \{D(g)\}$ is an IRR of dim σ, then

$$\sum_g D(g)_{ij}D(g^{-1})_{k\ell} = \frac{n}{\sigma}\delta_{i\ell}\delta_{jk}. \qquad (*)$$

If Γ is also unitary, then

$$\sum_g D(g)_{ij}D(g)^\dagger_{k\ell} = \frac{n}{\sigma}\delta_{i\ell}\delta_{jk}.$$

Theorem

If $\Gamma_i = \{D^{(i)}(g)\}$, $i = 1,\,2$ are inequivalent IRR's of dimensions σ_i, then

$$\sum_g D^{(1)}(g)_{ij}D^{(2)}(g^{-1})_{k\ell} = 0. \qquad (**)$$

Proof:

Let A be a matrix with σ_1 rows and σ_2 columns, and

$$P = \frac{1}{n}\sum_g D^{(1)}(g)AD^{(2)}(g^{-1}).$$

Then as before

$$D^{(1)}(s)P = PD^{(2)}(s).$$

By Schur's Lemma (since $D^{(i)}$ are IRR and inequivalent), $P = 0$. Now proceed as before, choosing a basis for $\sigma_1 \times \sigma_2$ matrices to find the result.

ii) Character theory

The character of g in a representation $\Gamma = \{D(g)\}$ is $\chi(g) \equiv \operatorname{tr} D(g)$. A character in an irreducible representation is "primitive", in a reducible representation it is "compound".

1) $\chi(g)$ is invariant under similarity transformations on Γ.

2) If G is finite, then $\chi(g^{-1}) = \chi^*(g)$.

For if $D(g)$ is unitary,

$$
\begin{aligned}
\chi(g^{-1}) &= \operatorname{tr} D(g^{-1}) \\
&= D(g)^\dagger \\
&= \operatorname{tr} D^*(g) \\
&= \chi^*(g).
\end{aligned}
$$

If Γ is not unitary, it is equivalent to a unitary representation for a finite G, so use 1), then above, to find the result.

If Γ is IRR, with associated character $\chi(g)$, set $i = j$, $k = \ell$ and sum on i, k in (*) to find

$$
\sum_g \chi(g)\chi(g^{-1}) = \frac{n}{\sigma} \cdot \sigma = n
$$

or

$$
\sum_g \chi(g)\chi^*(g) = n.
$$

If Γ_1 and Γ_2 are inequivalent IRR's, with associated characters $\chi^{(i)}(g)$, use (**) to find

$$
\sum_g \chi^{(1)}(g)\chi^{(2)}(g^{-1}) = \sum_g \chi^{(1)}(g)\chi^{(2)*}(g) = 0.
$$

Lemma

The characters in any representation (irreducible or not) are class functions on G, i.e., they have the same value in each class.

For

$$
\chi(sgs^{-1}) = \operatorname{tr} D(s)D(g)D(s^{-1}) = \chi(g).
$$

Lemma

The number of distinct inequivalent IRR's of a finite group G is less than or equal to the number of classes in G.

Proof:

Let $C_1, C_2, \ldots, C_p$ be the partition of G into classes C_i having n_i elements. Let $\Gamma^{(\alpha)}$ ($\alpha = 1, \ldots$) denote the inequivalent IRR's with characters χ_α. Then from above,

$$\sum_g \chi_\alpha(g)\chi_\beta(g)^* = n\delta_{\alpha\beta}.$$

If χ_α^k denotes the value of χ_α in class C_k, then

$$\sum_{k=1}^p n_k \chi_\alpha^k \chi_\beta^{k*} = n\delta_{\alpha\beta}.$$

Now call

$$v_\alpha^k = \left(\frac{n_k}{n}\right)^{\frac{1}{2}} \chi_\alpha^k.$$

Then we have

$$\sum_{k=1}^p v_\alpha^k v_\beta^{k*} = \delta_{\alpha\beta}.$$

Consequently, $v_\alpha \equiv (v_\alpha^1, \ldots, v_\alpha^p)$ are linearly independent. But v_α are p-dimensional vectors, so that there are at most p of them. Q.E.D.

Theorem

The number of distinct inequivalent IRR's is equal to the number of classes.

For proof, see the references for finite groups listed at the end of the lectures.

LECTURE 10

i) Character theory (continued)

Lemma

Two representations Γ, Γ' are equivalent if the associated characters χ, χ' are the same.

For $\Gamma = \oplus f_\alpha \Gamma_\alpha$ where:

$$f_\alpha \Gamma_\alpha = \Gamma_\alpha \oplus \Gamma_\alpha \oplus \ldots \qquad f_\alpha \text{ times,}$$

and where equivalent representations have been identified. Also, $\Gamma' = \oplus f'_\alpha \Gamma_\alpha$. Thus

$$\chi(g) = \sum_\alpha f_\alpha \chi_\alpha(g)$$

and

$$\chi'(g) = \sum_\alpha f'_\alpha \chi_\alpha(g).$$

Only if: Γ is equivalent to $\Gamma' \Rightarrow f_\alpha = f'_\alpha \Rightarrow \chi = \chi'$.

If: Assume $\chi = \chi'$, then $\sum_g \chi^*_\alpha \chi_\beta = n\,\delta_{\alpha\beta} \Rightarrow \chi_\alpha$ are linearly independent. So $\chi = \chi' \Rightarrow f_\alpha = f'_\alpha$.

The character in a representation completely fixes the representation up to equivalence. For

$$\chi(g) = \sum_\alpha f_\alpha \chi_\alpha(g)$$

$$\Rightarrow \sum_g \chi^*_\alpha(g)\chi(g) = n\, f_\alpha.$$

Therefore the multiplicity of IRR Γ_α in Γ is given by

$$f_\alpha = \frac{1}{n} \sum_g \chi^*{}_\alpha(g)\chi(g).$$

Since $\chi(g)$ fixes f_α, the result follows.

Frobenius criterion for irreducibility

A representation Γ with character χ is IRR iff

$$\sum_g \chi^*(g)\chi(g) = n = \text{order of group}.$$

For

$$\sum_g \chi^*(g)\chi(g) = n \sum_\alpha f_\alpha^2.$$

But Γ is IRR iff $\sum_\alpha f_\alpha^2 = 1$. Hence the result.

ii) The regular representation of a finite group

The regular representation introduced here is equivalent to the previous regular representation. (See Lecture 5.)

An algebra A is a vector space which is closed under a bilinear multiplication. [That is, for $x,\ y \in A$, $x \cdot y \equiv xy \in A$, and $(\alpha x + \beta y)z = \alpha(xz) + \beta(yz)$ for all complex numbers α, β. A similar property holds for the second factor.]

Examples:

1) Simple matrix algebra in n dimensions. The set of all $n \times n$ matrices forms an algebra under matrix multiplication. The multiplication satisfies the associativity,

$$(LJ)K = L(JK),\quad L, J, K \in \text{matrix algebra}.$$

Therefore, this is an associative algebra.

2) Angular momentum algebra. It is spanned by L_1, L_2, L_3 where

$$[L_i,\ L_j] = \epsilon_{ijk}\, L_k.$$

The algebra is the vector space $\{\sum_i \xi_i L_i\}_{\xi_i \in \mathbb{C}}$, and the composition is $L \cdot J \equiv [L, J]$ for L, $J \in$ algebra. [Here $\mathbb{C} \equiv$ set of complex numbers.] This is not an associative algebra. It is a Lie algebra.

A Lie algebra has the additional properties:

a) Antisymmetry: $L \cdot J = -J \cdot L$,

b) Jacobi Identity: $L \cdot (J \cdot K) + J \cdot (K \cdot L) + K \cdot (L \cdot J) = 0$.

iii) The group or Frobenius algebra $\mathbb{C}G$ for a finite group G

Let $G = \{g_i\}$. Take all formal linear combinations of the g_i's and make them into a vector space. This has a basis g_1, ..., g_n.

On this vector space, define bilinear composition by

$$\left(\sum_i \xi_i g_i \right) \left(\sum_j \eta_j g_j \right) = \sum_{ij} (\xi_i \eta_j)(g_i g_j)$$

where $(g_i g_j)$ is group composition. Then the space becomes the group algebra $\mathbb{C}G$.

Now we can define the regular representation of G by

$$G \ni s \to T(s) = \text{linear operator on } \mathbb{C}G$$

where

$$T(s) \sum_i \xi_i g_i = \sum_i \xi_i(sg_i).$$

Obviously

$$T(s_1)\, T(s_2) = T(s_1 s_2).$$

Note: We are essentially realizing G here as a subgroup of S_n.

Let us find the matrices $D(s)$ of $T(s)$ in the basis g_1, ..., g_n:

$$T(s)g_i = D(s)_{ji}\, g_j$$
$$= sg_i$$
$$= g_{i'} \text{ for some } i'.$$

This implies that

$$D(s)_{ji} = 1 \text{ if } j = i',$$
$$= 0 \text{ if } j \neq i'.$$

Also, when 1 is along the diagonal, $sg_i = g_i \Rightarrow s = e$. Thus tr $D(s) = \chi(s) = 0$ unless $s = e$, while $\chi(e) = n$.

As before, let Γ_α denote inequivalent IRR's with characters χ_α. Let dim $\Gamma_\alpha = \sigma_\alpha$.

The multiplicity of Γ_α in the regular representation is given by

$$\frac{1}{n}\sum_g \chi_\alpha^*(g)\chi(g) = \frac{1}{n}\chi_\alpha^*(e)\chi(e) = \frac{1}{n}\sigma_\alpha n = \sigma_\alpha.$$

So the IRR Γ_α of dim σ_α occurs σ_α times in the regular representation Γ_R, or

$$\Gamma_R = \oplus\, \sigma_\alpha \Gamma_\alpha.$$

This implies that by reducing the regular representation, we can find all irreducible representations.

Lemma

$$n = \sum_\alpha \sigma_\alpha^2$$

where the summation is over classes.

For

(a) number of classes = number of distinct IRR's,

(b) in a suitable basis,

$$D(s) = \begin{bmatrix} D^{(1)}(s) & & & & \\ & \ddots & & & \\ & & D^{(1)}(s) & & \\ & & & D^{(2)}(s) & \\ & & & & \ddots & \\ & & & & & D^{(2)}(s) & \\ & & & & & & \ddots \end{bmatrix}$$

In $D(s)$, each $D^{(i)}(s)$ is a square matrix of dimension $\sigma_i \times \sigma_i$ and occurs σ_i times. Q.E.D.

Examples:

1) S_2. There are two classes,

$$\{e\}, \quad \left\{ \begin{pmatrix} 1 \ 2 \\ 2 \ 1 \end{pmatrix} \right\}.$$

Applying the lemma with $n = 2$ gives

$$2 = \sigma_1^2 + \sigma_2^2 \Rightarrow \sigma_i = 1$$

and one has that there are two inequivalent one-dimensional IRR's.
They are given by

1) $e \to 1; \begin{pmatrix} 1 \ 2 \\ 2 \ 1 \end{pmatrix} \to 1.$

2) $e \to 1; \begin{pmatrix} 1 \ 2 \\ 2 \ 1 \end{pmatrix} \to -1.$

2) S_3. There are three classes, as we shall see later:

$$\{e\}, \quad \left\{ \begin{pmatrix} i \ j \ k \\ j \ i \ k \end{pmatrix} \mid i \neq j \neq k \right\}, \quad \left\{ \begin{pmatrix} i \ j \ k \\ j \ k \ i \end{pmatrix} \mid i \neq j \neq k \right\}.$$

We have

$$n = 6 = \sigma_1^2 + \sigma_2^2 + \sigma_3^2,$$

or

$$\sigma_1 = 1, \ \sigma_2 = 1, \ \sigma_3 = 2.$$

The two one-dimensional representations are given by

1) all elements of $S_3 \to 1$.

2) all even elements of $S_3 \to 1$, all odd elements of $S_3 \to -1$. The kernel in this map is the alternating subgroup A_3 of all even elements.

1) and 2) exist for all S_n. As for 2), recall that A_n is an invariant subgroup and $S_n/A_n = S_2$.

iv) The symmetric or permutation group S_n

Recall that

$$S_n \ni g = \begin{pmatrix} 1 \ \dots \ n \\ g_1 \ \dots \ g_n \end{pmatrix}, \quad g_i \text{ integers},$$

$$g^{-1} = \begin{pmatrix} g_1 \ \dots \ g_n \\ 1 \ \dots \ n \end{pmatrix}.$$

v) Cycles in S_n

A cycle of period or length two or a transposition is

$$\begin{pmatrix} i\ j\ k\ \ell\ \dots \\ j\ i\ k\ \ell\ \dots \end{pmatrix} \equiv (ij),\ i \neq j.$$

A cycle of period three is

$$\begin{pmatrix} i\ j\ k\ \ell\ \dots\ n \\ j\ i\ k\ \ell\ \dots\ n \end{pmatrix} = (i\ j\ k),\ i,\ j,\ k \text{ being unequal.}$$

Sometimes we add unary cycles in the notation:

$$\begin{pmatrix} i\ j\ \ell\ m\ \dots \\ j\ i\ \ell\ m\ \dots \end{pmatrix} = (ij)(\ell)(m)(n)\dots\ .$$

Any permutation

$$\begin{pmatrix} 1\ \dots\ n \\ s_1\ \dots\ s_n \end{pmatrix}$$

is a unique product of cycles, no two of which have any entry in common, provided the total number of entries in all cycles is n. (The last statement is to take care of the ambiguities in unaries.)

Proof:

Start with 1 on top, go to s_1 below. Start with s_1 on top, go to s_2 below, and so on. Eventually you come back to 1 below.

This completes a cycle. Start with any integer on top which is not included in the above process and repeat.

If s has α_1 cycles of length 1, α_2 cycles of length 2, ..., α_n cycles of length n, then

$$n = \sum_{i=1}^{n} i\alpha_i.$$

LECTURE 11

i) Cycles in S_n (continued)

Lemma

All elements $s \in S_n$ with the same cycle structure (i.e., same α_i) belong to the same class. If they have different cycle structures, they are in different classes.

Proof:

a) Let $s = s_1 \ldots s_\rho$, $s_i =$ cycles. Then $gsg^{-1} = gs_1g^{-1} \ldots gs_\rho g^{-1}$.

We show that gs_ig^{-1} is also a cycle of the same length as s_i. It follows that s, and gsg^{-1} have the same cycle structure.

Write

$$g = \begin{pmatrix} 1 \ldots & n \\ g_1 \ldots & g_n \end{pmatrix},$$

$$g^{-1} = \begin{pmatrix} g_1 \ldots & g_n \\ 1 \ldots & n \end{pmatrix}$$

and

$$s = \begin{pmatrix} 1 \ldots & n \\ s_1 \ldots & s_n \end{pmatrix}.$$

Then

$$gsg^{-1} = \begin{pmatrix} g_1 \ldots & g_n \\ g_{s_1} \ldots & g_{s_n} \end{pmatrix}.$$

Thus

$$g(a_1 \ldots a_k)g^{-1} = (g_{a_1}, g_{a_2}, \ldots g_{a_k}).$$

55

Or the cycle structure of gsg^{-1} is the same as the cycle structure of s.

b) Let $s = s_1 \ldots s_\rho$, $t = t_1 \ldots t_\rho$ where s_i, t_i are cycles of the same length, i.e., $s_i = (a_1^{(i)}, \ldots, a_\lambda^{(i)})$, $t_i = (b_1^{(i)}, \ldots, b_\lambda^{(i)})$. We must show that $\exists\, g$ such that $t = gsg^{-1}$. In the above, choose $g_{a_j^{(i)}} = b_j^{(i)}$. The result follows from the formula above.

Note: s and s^{-1} belong to the same class. For

$$(i_1 \ldots i_m)^{-1} = (i_m \ldots i_1)$$

is also a cycle of length m. Now if $s = s_1 \ldots s_\rho$ then $s^{-1} = s_\rho^{-1} \ldots s_1^{-1}$. Since each cycle preserves its length under inversion, the previous theorem shows that they are in the same class.

Let

$$\lambda_1 = \alpha_1 + \alpha_2 + \ldots + \alpha_n,$$
$$\lambda_2 = \phantom{\alpha_1 + {}} \alpha_2 + \ldots + \alpha_n,$$
$$\vdots \vdots$$
$$\lambda_n = \alpha_n$$

where $\lambda_1 \geq \lambda_2 \geq \ldots \geq \lambda_n \geq 0$, λ_i are integers and

$$\sum_{i=1}^{n} \lambda_i = \sum_{i=1}^{n} i\alpha_i = n.$$

By definition, $(\lambda_1, \ldots, \lambda_n)$ is a *partition* of n. It follows that the classes of S_n are in $1-1$ correspondence with the partitions of n.

Often, 0's are omitted in the notation, and repeated λ's are denoted by powers.

Examples:

S_8:

$(3,\, 2,\, 1,\, 1,\, 1,\, 0,\, 0,\, 0) \equiv (3,\, 2,\, 1^3)$.

S_3:

Three partitions : $(3,\, 0,\, 0)$, $(2,\, 1,\, 0)$, $(1,\, 1,\, 1)$.

$(3,\, 0,\, 0)$: 3 unaries($\lambda_1 - \lambda_2 =$ the number of unaries) $\Rightarrow$ class $= \{e\}$.

$(2,\, 1,\, 0)$: 1 unary, 1 cycle of length 2 $\Rightarrow$ class $= \left\{ \begin{pmatrix} i\ j\ k \\ i\ k\ j \end{pmatrix} \right\}$.

(1, 1, 1) : 1 cycle of length 3 $\Rightarrow$ class $= \left\{ \begin{pmatrix} i & j & k \\ j & k & i \end{pmatrix} \right\}$.

Transpositions

Any permutation is a product of transpositions. It is sufficient to prove this for cycles. Now

$$(s_1 s_3)(s_1 s_2) = (s_1 s_2 s_3), \quad (s_1 s_4)(s_1 s_2 s_3) = (s_1 s_2 s_3 s_4), \quad \text{etc.}$$

and therefore,

$$(s_1 \ \ldots \ s_k) = (s_1 s_k)(s_1 s_{k-1}) \ldots (s_1 s_2).$$

The way we write s as a product of transpositions is not unique, but the evenness or oddness of the number of transpositions is unique. [The *signature* ε_s of s is $+1$ if s is even and -1 if s is odd. Here s is even/odd if the number of transpositions is even/odd.]

Proof:

Allow s to act on n variables x_1, x_2, $\ldots$, x_n (in the natural way) as permutations. Let

$$\Delta \equiv \prod_{i<j}(x_i - x_j).$$

If $t = $ a transposition, then

$$t\Delta = -\Delta.$$

If s is arbitrary and can be written as a product of transpositions with an even or an odd number of factors, then

$$s\Delta = \Delta$$
$$= -\Delta,$$

a contradiction.

ii) Reduction of regular representation of S_n

Recall the following:
1) $\Gamma_\alpha = $ IRR of dim σ_α of S_n.
2) S_n acts by multiplication on the left on its algebra $\mathbb{C}S_n$.

3) $\mathbb{C}S_n = \oplus_\alpha \oplus_{j=1}^{\sigma_\alpha} \ell_\alpha^j$, dim $\ell_\alpha^j = \sigma_\alpha$.

Here ℓ_α^j for fixed α, different j carry equivalent representations. Thus

$$S_n \, \ell_\alpha^j = \ell_\alpha^j$$
$$(\Rightarrow \mathbb{C}S_n\ell_\alpha^j = \ell_\alpha^j$$
$$\Rightarrow \ell_\alpha^j \text{ is a "minimal left ideal"}).$$

If x is any fixed element in $\mathbb{C}S_n$, $\mathbb{C}S_n x$ is invariant under left action, i.e., $\mathbb{C}S_n\mathbb{C}S_n x = \mathbb{C}S_n x$. We will find suitable x's $\equiv e^j{}_\alpha$'s to reproduce ℓ_α^j's. Eventually we will have $e = \sum e_\alpha^j$. Then $\mathbb{C}S_n = \mathbb{C}S_n e = \otimes \mathbb{C}S_n e_\alpha^j$ where $\mathbb{C}S_n e_\alpha^j$ will be ℓ_α^j.

Definition. For a partition $(\lambda_1, \ldots, \lambda_n)$ a *Young frame* is defined as follows. If $\lambda_{k+1} = \ldots = \lambda_n = 0$, the frame is:

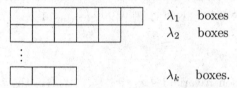

λ_1 boxes
λ_2 boxes
$\vdots$
λ_k boxes.

Definition. A *Young tableau* is a Young frame with 1, 2, $\ldots$, n arranged in the boxes in any order. There are $n!$ tableaux for a given frame.

Definition. A permutation p is *horizontal* for a tableau if it interchanges numbers only along the rows.

Example:

S_3: For $\begin{array}{|c|c|} \hline 2 & 1 \\ \hline 3 \\ \cline{1-1} \end{array}$,

$$\begin{pmatrix} 1\,2\,3 \\ 2\,1\,3 \end{pmatrix}$$

is a p, but

$$\begin{pmatrix} 1\,3\,2 \\ 3\,2\,1 \end{pmatrix}$$

is not a p.

Definition. A permutation q is *vertical* for a tableau if it interchanges numbers only along its columns.

Theorem

For a given tableau, let $P = \sum p$, $Q = \sum \varepsilon_q q$ (ε_q is the signature of q). Then $\mathbb{C}S_n PQ$ carries an irreducible representation of S_n.

Example:

S_n:

a) For the tableau $\boxed{1}\ \boxed{2}\ \boxed{\cdots}\ \boxed{n}$,

$$Q = e,$$
$$P = \sum_{s \in S_n} s.$$

For $s' \in S_n$, $s'PQ = s'P = P$. This implies that $\mathbb{C}S_n PQ$ is one-dimensional, with basis PQ. The matrix of any $s' \in S_n$ is 1. This is the trivial representation.

b) For the tableau $\begin{array}{|c|} \hline 1 \\ \hline 2 \\ \hline \vdots \\ \hline n \\ \hline \end{array}$,

$$P = e,$$
$$Q = \sum \varepsilon_s s.$$

Then $s'Q = \sum \varepsilon_s s's = \varepsilon_{s'} \sum \varepsilon_{s's}(s's) = \varepsilon_{s'}Q$. Therefore $\mathbb{C}S_n PQ$ is one-dimensional. The matrix of s' is equal to $\varepsilon_{s'}$.

Note: All tableaux with one row give identical vector spaces. All tableaux with one column give identical vector spaces. PQ's arising from different frames lead to inequivalent IRR's. Those arising from the same frame, but different tableaux give equivalent IRR's. In fact it is possible that for two such PQ's, say $P_1 Q_1$ and $P_2 Q_2$, $\mathbb{C}S_n P_1 Q_1 \cap \mathbb{C}S_n P_2 Q_2$ is not $\{0\}$.

[For the proofs of these results and those that follow, the references cited in the bibliography may be consulted.]

Now we give a prescription for a) finding ℓ^j_α and b) finding a basis for ℓ^j_α. From b) we can find matrices of IRR by computation. (Let $\{g\}$ act on the basis elements.)

Definition. For a given frame, T is a standard tableau if the numbers along each row increase from left to right, and increase along each column from top to bottom.

Example:

S_3:

IRR	*dimension*	*standard*	*non-standard*

1   dimension   standard   non-standard

1

| 1 | 2 | 3 |

anything else,

| 3 | 1 | 2 |

etc.

1

1
2
3

2
1
3

etc.

2

1	2
3	

,

1	3
2	

2	1
3	

Theorem

The number of standard tableaux for a given frame is the dimension of the IRR associated with a given frame. (See example above.)

See Boerner for a formula for the dimension.

Theorem

(Construction of ℓ_α^j and its basis).

For a given frame, let $T(\{\lambda_i\})_1,\ T(\{\lambda_i\})_2,\ \ldots,\ T(\{\lambda_i\})_d$ denote the distinct standard tableaux (d was called σ_α before). We fix the labelling in what follows. Let $\pi_j \in S_n$ be the permutation such that

$$\pi_j T(\{\lambda_i\})_1 = T(\{\lambda_i\})_j.$$

Let $P = \sum p,\ Q = \sum \varepsilon_q q$ for tableau 1, i.e., $T(\{\lambda_i\})_1$. Define

$$Y(\lambda_i)_1 = \frac{d}{n!} PQ.$$

Then

$$[Y(\lambda_i)_1]^2 = Y(\lambda_i)_1$$

and

$$\pi_j Y(\lambda_i)_1, \; j \; = \; 1, \; \ldots, \; d$$

is a basis for $\mathbb{C} S_n Y(\lambda_i)_1$.

The process can be represented by regarding tableau $T(\lambda_i)_j \; [j \; = \; 2, \; 3,$ $\ldots, \; d]$ as the first tableau. Let $Y(\lambda_i)_j$ be the associated $\frac{d}{n!} PQ$. Then $\mathbb{C} S_n Y(\lambda_i)_j \cap \mathbb{C} S_n Y(\lambda_i)_k = \{0\}$ for $j \neq k$.

Thus all ℓ_α^j have been identified.

Remarks:

1) The PQ for a given tableau is called a Young symmetrizer. Also $(PQ)^2$ = constant (PQ).

2) We can use QP instead of PQ, the result is that IRR's undergo a similarity transformation.

3) The matrices of the IRR's constructed as above are equivalent to unitary ones.

4) It is possible to write the identity as a linear combination of Young symmetrizers of standard tableaux.

5) S_n is generated by the two elements (12), $(123 \ldots n)$.

Proof:

Let $a = (12)$, $b = (12 \ldots n)$. Every permutation is a product of transpositions. Now $bab^{-1} = (23)$, $b(23)b^{-1} = (34)$, etc. (see problem set). Thus all transpositions of the form $(k \; k + 1)$ are generated by this procedure. But for $i < k$,

$$(k \; k + 1)(i \; k)(k \; k + 1) = \begin{pmatrix} i & k+1 & k \\ k+1 & i & k \end{pmatrix}$$
$$= (i \; k + 1).$$

So all transpositions are generated by a and b (e.g. take $(ik) = (34)$. Then one gets (35), (36),..., $(3n)$ by the above procedure). Q.E.D.

It follows that it is sufficient to find the matrices of (12) and $(12 \ldots n)$ in an IRR by the preceding method. The rest of the matrices can be found from these two.

PART 3
LIE GROUPS

LECTURE 12

i) Definition and examples

Definition. A group G is an n-parameter continuous group if its elements can be parametrized by n real variables varying in a continuous range and if no more than n such parameters are necessary.

Call n the dimension of G. [This is a new definition of dimension.] Write $G = \{g(a)\}$, $a = (a_1, \ldots, a_n)$. The multiplication rule is given by

$$g(a)g(b) = g[\phi(a,b)],$$
$$\phi = (\phi_1, \ldots, \phi_n).$$

Normally we choose the a_i so that $a_i = 0$ gives the identity e.

Lie Group: For a Lie group the power series expansion of ϕ in a and b around any two points a_0 and b_0 converges to ϕ, i.e., ϕ is an analytic function of a and b.

Examples:

1) $GL(n,\mathbb{C})$: The general linear group of complex nonsingular $n \times n$ matrices. Dimension $= 2n^2$.

2) $GL(n,\mathbb{R})$: The subgroup of real matrices of $GL(n,\mathbb{C})$. Dimension $= n^2$.

3) $SL(n,\mathbb{C})$: The complex unimodular group in n dimensions. It is the subgroup of $GL(n,\mathbb{C})$ of matrices of determinant 1. Dimension $= 2n^2 - 2$. (If $\det g = 1$ for $g \in GL(n,\mathbb{C})$, one has two real constraint equations, Re $\det g = 1$ and Im $\det g = 0$.)

4) $SL(n,\mathbb{R})$: The subgroup of $SL(n,\mathbb{C})$ of real matrices. Dimension $= n^2 - 1$.

5) $U(n)$: The unitary group in n dimensions. It is the subgroup of $GL(n, \mathbb{C})$ made up of unitary matrices. $u \in U(n)$ means (a) different columns of u are orthogonal,

$$\sum_i u_{ij}^* u_{ik} = 0, \quad j \neq k;$$

b) each column is normalized to unity,

$$\sum_i u_{ij}^* u_{ij} = 1.$$

To compute the dimension of $U(n)$ note that (a) gives $n-1$ complex constraint equations for the first row, $n-2$ equations for the second row, etc., implying $n(n-1)/2$ complex equations or $n(n-1)$ real equations, and (b) gives n real equations. So dimension of $U(n) = 2n^2 - (n(n-1) + n) = n^2$.

6) $SU(n)$: The special unitary group in n dimensions. It is the subgroup of $U(n)$ of matrices with determinant 1. Dimension $= n^2 - 1$. (If $u \in U(n)$, then $\det u = e^{i\alpha}$, α real. So we need only one equation, say Re $\det u = 1$, to get $\det u = 1$.)

7) $O(n, \mathbb{C})$: The complex orthogonal group. $g \in O(n, \mathbb{C})$ means that $g^T g = 1$ where g is an $n \times n$ complex matrix. Now the orthogonality of rows gives $n(n-1)$ real conditions, while the normalization of each row gives $2n$ real conditions. So dimension of $O(n, \mathbb{C}) = 2n^2 - (n(n-1) + 2n) = n^2 - n = n(n-1)$.

8) $O(n, \mathbb{R}) \equiv O(n)$: The real orthogonal group. It is the real subgroup of $O(n, \mathbb{C})$. Dimension of $O(n) = n(n-1)/2$.

9) $SO(n, \mathbb{C})$, $SO(n, \mathbb{R})$: The proper subgroups of $O(n, \mathbb{C})$ and $O(n)$. If $g \in O(n, \mathbb{C})$, $\det g = \pm 1$, so select that part of $O(n, \mathbb{C})$ with $\det g = +1$ to get $SO(n, \mathbb{C})$. This does not change the dimension. $SO(n, \mathbb{R}) \equiv SO(n)$ is found in a similar way.

Many of these groups leave some quadratic form invariant.

Examples:

(1) $U(n)$ or $SU(n)$ leaves

$$(x, y) \equiv \sum_{i=1}^n x_i^* y_i$$

invariant. One can define $U(n)$ by this requirement.

(2) $O(n, \mathbb{C})$, $O(n)$, $SO(n, \mathbb{C})$, $SO(n)$ leave

$$(x, \ y) \equiv \sum_{i=1}^n x_i y_i$$

invariant.

By considering other quadratic forms, we get other groups as indicated below. The notation will indicate the quadratic form.

10) $U(p,q)$: The form which is invariant under $U(p,q)$ is

$$\sum_{i=1}^{p} x_i^* y_i - \sum_{i=p+1}^{p+q} x_i^* y_i$$

i.e. if $u \in U(p,q)$.

$$u^\dagger M u = M$$

where

$$M = \begin{bmatrix} 1_{p \times p} & 0_{p \times q} \\ 0_{q \times p} & -1_{q \times q} \end{bmatrix}.$$

$SU(p,q)$ is the subgroup of $U(p,q)$ with det $= 1$.

11) $O(p,q|\mathbb{C}) : g \in O(p,q|\mathbb{C})$ leaves M above invariant in the sense that

$$g^T M g = M.$$

$O(p,q|\mathbb{R})$ is defined similarly, it being the real subgroup of $O(p,q|\mathbb{C})$. $O(3,1|\mathbb{R}) \equiv O(3,1)$ leaves the form

$$\sum_{i=1}^{3} x_i y_i - x_0 y_0$$

invariant. $O(3,1)$ is the full Lorentz group with dimension $n(n-1)/2 = 6$. If $\Lambda \in O(3,1)$, then det $\Lambda = \pm 1$. $SO(3,1)$ is the subgroup of $O(3,1)$ with det $= +1$. The set of $\Lambda \in SO(3,1)$ with Λ_{00} positive (i.e., these $\{\Lambda\}$ do not change the sign of time) is called the proper orthochronous Lorentz group $\mathcal{L}_+^\uparrow$.

12) The conformal group is essentially $SO(4,2)$ or $SU(2,2)$. The de Sitter group is $SO(4,1)$.

13) $Sp(n,\mathbb{C})$, $Sp(n,\mathbb{R})$ are the symplectic groups. If $s \in Sp(n,\mathbb{C})$, it leaves

$$\varepsilon = \begin{bmatrix} 0_{n \times n} & 1_{n \times n} \\ -1_{n \times n} & 0_{n \times n} \end{bmatrix}$$

invariant, i.e.,

$$s^T \varepsilon s = \varepsilon.$$

If $s \in Sp(n,\mathbb{R})$, it is also real.

In physics, we encounter Lie groups which are not groups of matrices in their origin as the following examples show.

14) The group $\mathcal{T}_n$ of translations in n dimensions, i.e., on $\mathbb{R}^n$.

15) The group $\mathcal{E}_3$ of Euclidean motions in three dimensions.

16) The Galilei group. It is the group of transformations on x_1, x_2, x_3, t generated by

(a) rotations and reflections on $\vec{x}$,

(b) space and time translations,

(c) Newtonian velocity transformations : $\vec{x}' = \vec{x} - \vec{v}t$, $t' = t$.

17) The Poincaré group $\mathcal{P}$. As we saw, this is generated by $O(3,1)$ and translations on $\mathbb{R}^4$.

Finally we have

18) The group of canonical transformations in Classical Mechanics, the group of coordinate transformations in General Relativity and the group of gauge transformations in gauge theories. These groups are not of the categories considered so far. An element of these groups is labelled by a function or functions. It cannot be labelled by n numbers. Thus these groups are not in the category of continuous groups defied above.

Remarks:

It is possible to consider $GL(n, \mathbb{R})$ as a subgroup of $SL(n+1, \mathbb{R})$ using the isomorphism

$$GL(n, \mathbb{R}) \ni A \rightarrow \begin{bmatrix} A & 0 \\ 0 & \det A^{-1} \end{bmatrix} \in SL(n+1, \mathbb{R}).$$

Similarly the traslation group $\mathcal{T}_n$ can be identified as a subgroup of $GL(n+1, \mathbb{R})$ using the isomorphism

$$\mathcal{T}_n \ni A = (a_1, a_2, \cdots, a_n) \rightarrow \begin{bmatrix} 1 & \cdots & 0 & a_1 \\ \vdots & \ddots & \vdots & \vdots \\ 0 & \cdots & 1 & a_n \\ 0 & \cdots & 0 & 1 \end{bmatrix} \in GL(n+1, \mathbb{R}).$$

By using this trick, we can also consider the Euclidean, Galilei and Poincaré groups as groups of matrices.

LECTURE 13

i) More on $GL(n, \mathbb{C})$ and its subgroups

Most of the groups dealt with in this book arise as subgroups of the general linear group. We recall that this group is given by those linear transformations $L : V \to V$ which are invertible. Composition of maps

$$\phi(L_1, L_2) = L_2 \circ L_1,$$

i.e. $V \xrightarrow{L_2} V \xrightarrow{L_1} V$ gives the binary operation defining the group as from Lecture 1. When we fix a basis in V, with any linear transformation, we can associate a matrix and the above product is realized in terms of row-by-column product of matrices. The number of parameters needed to identify an element is clearly n^2 and a neighborhood of each element is simply a neighborhood of the point as an element of $\mathbb{R}^{n^2}$. The composition law $\phi : GL(n, \mathbb{C}) \times GL(n, \mathbb{C}) \to GL(n, \mathbb{C})$, where $GL(n, \mathbb{C})$ represents the set of invertible matrices, is a smooth map because $\phi(x, y) = z$ is a polynomial in the matrix elements x_{ij}, y_{ij}. Also the map associating any element $x \in GL(n, \mathbb{C})$ with its inverse, say x^{-1}, is smooth as well.

   This group is the prototype of a Lie group, i.e. a group space for which the composition and inverse maps are both smooth maps. We notice that the determinant map

$$\det : A \to \det A$$

is a polynomial map and therefore continuous. The group $GL(n, \mathbb{C}) \subset \mathbb{R}^{n^2}$ is identified as the complement of the closed subset $\{\det^{-1}(0)\}$, i.e. those matrices whose determinant is different from zero. Thus it is an open subset of $\mathbb{R}^{n^2}$. We can consider the differential calculus on it as we do for the vector space $\mathbb{R}^{n^2}$, because differentiability only requires local properties.

If $b : V \times V \to \mathbb{R}$ is any bilinear map, we define subgroups of the general linear group $GL(n, \mathbb{C})$ as the subset of transformations satisfying the property

$$b(Av_1, Av_2) = b(v_1, v_2).$$

The subset defined by this requirement is a subgroup because the composition $A_2 \circ A_1$ satisfies

$$b((A_2 \circ A_1)v_1, (A_2 \circ A_1)v_2) = b(A_2(A_1v_1), A_2(A_1v_2))$$
$$= b(A_1v_1, A_1v_2)$$
$$= b(v_1, v_2).$$

Because this subgroup is defined as the set of elements preserving the bilinear map b, it turns out to be a closed subgroup and by a deep theorem of E. Cartan, it is also a submanifold of $\mathbb{R}^{n^2}$.

Examples:

$V \equiv \mathbb{R}^2$, $v_1 = \begin{pmatrix} x_1 \\ x_2 \end{pmatrix}$, $v_2 = \begin{pmatrix} y_1 \\ y_2 \end{pmatrix}$. Define $b(v_1, v_2) = x_1 y_1 + x_2 y_2$. Then on the generic element

$$A = \begin{pmatrix} a_1 & a_2 \\ a_3 & a_4 \end{pmatrix}$$

with a_1, a_2, a_3, a_4 real numbers, we find the constraints arising from $b(Av_1, Av_2) = b(v_1, v_2)$ to be

$$a_1^2 + a_3^2 = 1, \quad a_2^2 + a_4^2 = 1, \quad a_1 a_2 + a_3 a_4 = 0.$$

It is not difficult to show that the only solution of these equations has the form

$$a_1 = \cos\alpha, \quad a_2 = \pm\sin\alpha, \quad a_3 = \mp\sin\alpha, \quad a_4 = \cos\alpha.$$

Thus $A = \begin{pmatrix} \cos\alpha & \sin\alpha \\ -\sin\alpha & \cos\alpha \end{pmatrix}$ is a one-parameter subgroup of the four-dimensional group $GL(2, \mathbb{R})$.

Similarly for $b(v_1, v_2) = x_1 y_1 - x_2 y_2$, we find the subgroup described by matrices

$$A = \begin{pmatrix} \cosh\alpha & \sinh\alpha \\ \sinh\alpha & \cosh\alpha \end{pmatrix}.$$

For $b(v_1, v_2) = x_1 y_2 - x_2 y_1$, imposing the condition $b(Av_1, Av_2) = b(v_1, v_2)$ we find the restriction on the four parameters given by

$$a_1 a_4 - a_2 a_3 = 1.$$

It is a three-parameter subgroup. It contains both the previous subgroups along with

$$A = \begin{pmatrix} e^\alpha & 0 \\ 0 & e^{-\alpha} \end{pmatrix}.$$

Another example is provided by $V = \mathbb{C}^2$. A generic element of $GL(2, \mathbb{C})$ is given by

$$A = \begin{pmatrix} a_1 & a_2 \\ a_3 & a_4 \end{pmatrix}$$

with $a_1, a_2, a_3, a_4 \in \mathbb{C}$. If we consider $b : V \times V \to \mathbb{C}$, defined by

$$b(v_1, v_2) = x_1^* y_1 + x_2^* y_2$$

(it is called a sesquilinear map because it is antilinear in the first vector), then matrices satisfying

$$b(Av_1, Av_2) = b(v_1, v_2)$$

define the unitary group $U(2) \subset GL(2, \mathbb{C})$. By imposing the additional requirement $\det A = 1$, we get the special unitary group $SU(2) \subset U(2)$. It is not difficult to see that the previous conditions amount in this case to

$$a_1^* a_1 + a_2^* a_2 = 1, \quad a_3^* a_3 + a_4^* a_4 = 1, \quad a_1^* a_2 + a_3^* a_4 = 0,$$

and the space of free parameters is the three-dimensional sphere $S^3 \subset \mathbb{R}^4 \equiv \mathbb{C}^2$. The composition map $\phi : S^3 \times S^3 \to S^3$ is a smooth map in both arguments, and also $s \to s^{-1}$, the inverse map, is smooth.

Previous examples may be generalized from a two-dimensional vector space to any n-dimensional vector space.

LECTURE 14

i) Continuous curves or paths in G

A map from $\mathbb{R}^1$ to G which is continuous is called a path or a curve in G. Thus

$$\{g[a(t)] \mid t \in \mathbb{R}^1;\ a(t) \text{ continuous in } t\}$$

is a path in G.

Two elements g_1 and $g_2 \in G$ are connected if they can be joined by a path, i.e., if $\exists\ g(t)$ continuous in t such that $g(0) = g_1$ and $g(1) = g_2$.

Examples:

$O(3)$: Let g, h be in $O(3)$ with $\det g = 1$, $\det h = -1$. Then g and h are not connected. For let

$$t \xrightarrow{\ s\ } s(t) = \{s_{ij}(t)\}$$

be a path connecting g to h. Since $s_{ij}(t)$ is continuous, $\det s(t)$ is continuous and therefore a constant. We thus have a contradiction.

ii) Connected component to identity G_0 of a Lie group G

Let

$$U = \{g(a) \in G \mid |a_i| < \varepsilon,\ \varepsilon > 0\}$$

such that $g \in U \Rightarrow g^{-1} \in U$. Such a U is called a group germ by Weyl.

Now consider

$$U^1 \equiv U,\ U^2 \equiv UU,\ U^3 \equiv UUU,\ \ldots$$

Call their union G_0. Then G_0 is a group. For

(a) If $s_k \in U^k$, $s_\ell \in U^\ell$, then $s_k s_\ell \in U^{k+\ell} \subset G_0$.

(b) $e \in G_0$.

(c) If $g \in G_0$, then $g = g_1 \ldots g_k$ for some k where $g_i \in U$. Thus $g^{-1} = g_k^{-1} \ldots g_1^{-1}$. But $g_i \in U$ implies that $g^{-1} \in U$. So $g^{-1} \in U^k \subset G_0$.

Pontrjagin in his book proves that G_0 does not depend on the choice of U.

We now show that G_0 is connected, that is, any two elements in G_0 are connected. If every s in G_0 is connected to e, then any s and s' in G_0 are connected by the path that goes first from s to e and then from e to s'. Thus it is sufficient to show that every s in G_0 is connected to e. By hypothesis, if $g(a)$ is in U, $g(ta)$ is in U for $0 \leq t \leq 1$, so that every element in U is connected to e. Now any element in G_0 can be written in the form

$$g(a)\ g(b)\ g(c)\ \ldots, \quad g(a),\ g(b),\ g(c),\ \ldots \in U.$$

The path $s(t)$ defined by

$$s(t) = g(ta)g(tb)g(tc)\ldots, \quad 0 \leq t \leq 1$$

then connects e to $g(a)\ g(b)\ g(c) \ldots$.

Examples:

1) $G_0 = U(1) = \{e^{i\theta}\}$. A choice for U is

$$U = \{e^{ia} \mid |a| < \varepsilon\}.$$

2) $G_0 = SO(3)$. Recall that every rotation is a rotation about an axis $\hat{n}$ by an angle θ. But a rotation by π around $\hat{n}$ is the same as a rotation by π around $-\hat{n}$.

Consider a ball B_3 in $\mathbb{R}^3$ of radius π around the origin O. A point P in this ball defines an axis given by the direction of OP and an angle given by the length of OP. By the preceding remark, this ball represents $SO(3)$ as a manifold if we also identify diametrically opposed points on its boundary. A choice for U is

$$U = \{\text{set of all P} \mid \text{length of OP} < \varepsilon\}.$$

That $SO(3)$ is the union of U, UU, $\ldots$ is manifest. One picks an axis and takes multiples of rotations along that axis contained in U until one reaches the boundary. Next one does this for all possible choices of the axis.

iii) Invariant integration

For finite groups, we saw that the sum

$$\sum_g f(g)$$

with the properties

$$\sum_g f(g) = \sum_g f(sg) \quad \text{(left invariance)}$$

$$= \sum_g f(gs) \quad \text{(right invariance)}$$

played an important role.

For Lie groups, there are two integration procedures leading respectively to left and right invariance. The left and right invariant (Haar or Hurwitz) measures are denoted by

$$d\mu_L(g) = \rho_L(a)d^n a,$$
$$d\mu_R(g) = \rho_R(a)d^n a, \quad g \equiv g(a).$$

Let $M \subset G$. Then the defining property of $d\mu_L$ is

$$\int_M d\mu_L(g)f(g) = \int_{g_0 M} d\mu_L(g)f(g_0^{-1}g),$$

i.e., the Jacobian of the transformation $g \to g_0 g$ is 1. Equivalently $d\mu_L(g_0 g) = d\mu_L(g)$.

Similarly, for the right invariant integral,

$$\int_M d\mu_R(g)f(g) = \int_{M g_0} d\mu_R(g)f(gg_0^{-1})$$

or $d\mu_R(g) = d\mu_R(gg_0)$.

If the domain of integration M is G, then

$$\int_G d\mu_L(g)f(g) = \int_G d\mu_L(g)f(g_0^{-1}g)$$

and likewise,

$$\int_G d\mu_R(g)f(g) = \int_G d\mu_R(g)f(gg_0^{-1}).$$

Claim

$$\rho_L(a) = \left[\det\left(\frac{\partial \phi^\rho(a,b)}{\partial b^\sigma} \right) \Big|_{b=0} \right]^{-1},$$

$$\rho_R(a) = \left[det\left(\frac{\partial\phi^\rho(b,a)}{\partial b^\sigma}\right)\Big|_{b=0}\right]^{-1}.$$

The proof of this result is one of the problems.

Remarks:

1) These measures are unique up to multiplicative constants.

2) A Lie group is compact iff

$$V \equiv \int_G d\mu_L(g) \quad \text{exists}.$$

Then $d\mu_R = d\mu_L$ up to a constant. V is called the volume of the group. (In the above one can of course interchange the roles of R and L.)

3) $d\mu_L = $ constant $\cdot d\mu_R$ for simple and Abelian groups.

4) The measures $d\mu_L$, $d\mu_R$ are actually measures (after the choice of an overall sign), so that, if $f \geq 0$ on $M \subset G$, then

$$\int_M d\mu_{L,R} \, f \geq 0,$$

and its vanishing implies that $f = 0$ on M (almost everywhere).

So we can construct two Hilbert spaces using functions from G to $\mathbb{C}$:

$$H_L \;:\; \text{if } f_1, f_2 \in H_L, \text{ then}$$
$$(f_1, f_2)_L = \int_G d\mu_L \; f_1(g)^* f_2(g) < \infty.$$

$$H_R : \text{If } f_1, f_2 \in H_R, \text{ then}$$
$$(f_1, f_2)_R = \int_G d\mu_R \; f_1(g)^* f_2(g) < \infty.$$

The left regular representation is unitary in H_L and the right regular representation is unitary in H_R (as is easily proved).

Examples:

1) $G = $ the Lorentz group in 1+1 dimensions (one space and one time). This group describes Lorentz transformations along a line. We have

$$G = \left\{\begin{bmatrix} \cosh\theta & \cosh\theta \\ \sinh\theta & \sinh\theta \end{bmatrix}\right\}$$

where $-\infty < \theta < \infty$. θ is called the rapidity variable. Thus we can write $g = g(\theta)$. Since $g(\theta_1)g(\theta_2) = g(\theta_1 + \theta_2)$,

$$\phi(\theta_1, \theta_2) = \theta_1 + \theta_2$$

and

$$d\mu_L(g) = d\theta$$
$$= d\mu_R.$$

Thus

$$\int d\mu_L = \infty = \int d\mu_R$$

or this group is not compact.

2) $G = SO(3)$. The following argument is not rigorous, but it should make the answer plausible.

If R is in $SO(3)$, we can write $R = R_3(\phi)R_2(\theta)R_3(\gamma)$ where $R_i(\psi)$ is rotation by ψ around the ith axis. Now the map

$$R \xrightarrow{\ n\ } \hat{n}(R)$$

$$\equiv R \begin{pmatrix} 0 \\ 0 \\ 1 \end{pmatrix}$$

$$= R_3(\phi)R_2(\theta) \begin{pmatrix} 0 \\ 0 \\ 1 \end{pmatrix}$$

is a map from $SO(3)$ to the two-sphere S^2 parametrized by the Euler angles (θ, ϕ). It is compatible with the $SO(3)$ group actions $R \to SR$ and $\hat{n}(R) \to S \hat{n}(R)[S \in SO(3)]$, i.e.,

$$\hat{n}(SR) = S \hat{n}(R).$$

Now S^2 has the rotationally invariant measure $d\phi \, d\cos\theta$. Further as compared to S^2, $SO(3)$ contains in addition the degree of freedom of the Abelian subgroup $\{R_3(\gamma)\}$ whose invariant measure is of course $d\gamma$. [See previous example.] This suggests that for $SO(3)$,

$$d\mu_L(g) = d\phi \, d\cos\theta \, d\gamma.$$

The answer is correct. The group volume is

$$V = \int_0^{2\pi} d\phi \int_{-1}^{1} d\cos\theta \int_0^{2\pi} d\gamma$$
$$= 8\pi^2.$$

Hence $SO(3)$ is compact and $d\mu_L(R) = d\mu_R(R)$.

LECTURE 15

i) Invariant integration (continued)

Call $d\mu$ the invariant measure $d\mu_L = d\mu_R$ for compact groups.

A continuous function on a compact set is (uniformly) bounded. [$f(g)$ is continuous in $g \in G$ if $f[g(a)]$ is continuous in a]. We will consider only continuous representations $\{D(g)\}$ where each matrix element $D_{ij}(g)$ is continuous in g. Then the various integrals we write in the statements below exist for compact groups. These statements are proved as in the case of finite groups.

1) Every finite dimensional representation of a compact Lie group is equivalent to a unitary one, and hence completely reducible to a direct sum of irreducible representations.

2) Let G be a compact Lie group with distinct finite dimensional IRR's $\Gamma^{(\alpha)} = \{D^{(\alpha)}\}$ where dim $\Gamma^{(\alpha)} = \sigma_\alpha$. Then

$$\int_G d\mu(g) D_{ij}^{(\alpha)}(g) D_{k\ell}^{(\beta)}(g^{-1}) = \delta_{\alpha\beta}\, \delta_{i\ell}\, \delta_{jk} \frac{V}{\sigma_\alpha}$$

where

$$V = \int_G d\mu(g).$$

3) If $\chi^{(\alpha)}$ is the character in $\Gamma^{(\alpha)}$ ($\chi^{(\alpha)}(g) = \mathrm{Tr}\, D^{(\alpha)}(g)$),

$$\chi^{(\alpha)}(g^{-1}) = \chi^{(\alpha)}(g)^*,$$
$$\int_G d\mu(g) \chi^{(\alpha)}(g)^* \chi^{\beta}(g) = V\delta_{\alpha\beta}.$$

4) If $\Gamma = \bigoplus_\alpha f_\alpha \Gamma^{(\alpha)}$ where f_α is the multiplicity of $\Gamma^{(\alpha)}$ in Γ, and if χ is the character in Γ, then

$$f_\alpha = \frac{1}{V} \int d\mu(g) \chi^{(\alpha)}(g)^* \chi(g).$$

5) *The Frobenius Criterion.*

Γ is an irreducible representation iff

$$\int_G d\mu(g)\chi^*(g)\chi(g) = V.$$

Examples:

1) $G = U(1) = \{e^{i\theta} \mid 0 \le \theta \le \pi\}$. The map $e^{i\theta} \to D^{(n)}(\theta) = e^{in\theta}$ defines for every integer n a distinct IRR $\Gamma^{(n)}$. Also $d\mu = d\theta$ and $V = 2\pi$. Thus, by above,

$$\int_0^{2\pi} d\theta D^{(n)}(\theta)^* D^{(m)}(\theta) = 2\pi\,\delta_{mn}.$$

This result can be trivially verified by direct integration.

2) $G = SO(3)$. If we write $R = R_3(\gamma)\,R_2(\beta)\,R_3(\alpha)$, then

$$d\mu = d\gamma \; d\cos\beta \; d\alpha, \quad V = 8\pi^2.$$

The IRR's are $\Gamma^{(j)} = \{D^{(j)}\}$, $j = 0, 1, 2, \ldots$ with $\sigma_1 = 2j + 1$. [For $SU(2)$, j can also be $1/2, 3/2, \ldots$.] Now

$$D^{(j)}(R) = e^{iJ_3\gamma}\, e^{iJ_2\beta}\, e^{iJ_3\alpha}$$

where J_i are the angular momentum operators in the spin j representation. In the conventional basis $\{|jm\rangle\}$ $[J_3|jm\rangle = m|jm\rangle, \; \langle jm'|jm\rangle = \delta_{m'm}]$,

$$D^{(j)}_{mn}(\gamma, \beta, \alpha) \equiv \langle jm|D^{(j)}|jn\rangle$$
$$= e^{im\gamma}\langle jm|e^{iJ_2\beta}|jn\rangle e^{in\alpha},$$
$$= e^{im\gamma}d^j{}_{mn}(\beta)e^{in\alpha}.$$

$d^j(\beta)$ is called the reduced rotation matrix. It is real for the standard phase choice. From

$$e^{iJ_2\beta_1}\, e^{iJ_2\beta_2} = e^{iJ_2(\beta_1+\beta_2)},$$

we have

$$d^j(\beta_1)d^j(\beta_2) = d^j(\beta_1 + \beta_2).$$

So

$$d^j(\beta)^{-1} = d^j(-\beta).$$

The orthogonality relation thus leads to

$$\int_{-1}^1 d\cos\beta \; d^{j_1}_{\rho\sigma}(\beta) \; d^{j_2}_{\sigma\rho}(-\beta) = \frac{2}{2j_1 + 1}\delta_{j_1 j_2}.$$

Textbooks on quantum mechanics give d^j in terms of Jacobi polynomials. Using this identity, we can rewrite the above equation as the conventional orthogonality relation of Jacobi polynomials.

ii) The left regular representation Γ^L on H_L

We assume that G is compact. Then we have the following results. [cf. Naimark or Talman.] [The representations are assumed to be continuous.]

1) Every finite dimensional IRR occurs in Γ^L. For [with fixed j], $D_{ij}^{(\alpha)*}$, $i = 1, 2, \ldots, \sigma_\alpha$ are functions from G to $\mathcal{C}$ and so span a subspace of H_L. It carries the representation $\Gamma^{(\alpha)}$:

$$D_{ij}^{(\alpha)}(g_0^{-1}g)^* = D_{i'i}^{(\alpha)}(g_0)\, D_{i'j}^{(\alpha)}(g)^*.$$

[Here, we have assumed that $D^{(\alpha)}$ is unitary. This is permissible as remarked earlier. Recall also that the functions $\{D_{ij}^{(\alpha)*}\}$ form an orthogonal system by previous results.]

2) *The Peter — Weyl Theorem.*

If $f \in H_L$,

$$f(g) = \Sigma \xi_{ij}^\alpha \, D_{ij}^{(\alpha)}(g)^*$$

[convergence being in the norm].

[Or equally well, $f(g) = \sum \eta_{ij}^\alpha \, D_{ij}^{(\alpha)}(g)$.]

Thus

$$H_L = \bigoplus_\alpha \bigoplus_{j=1}^{\sigma_\alpha} H_j^{(\alpha)}$$

where $H_j^{(\alpha)}$ for fixed α, different j, carry the same IRR $\Gamma^{(\alpha)}$. This means

$$\Gamma^L = \bigoplus_\alpha \sigma_\alpha\, \Gamma^{(\alpha)}$$

or the multiplicity of the IRR $\Gamma^{(\alpha)}$ in Γ^L is equal to its dimension.

The following can also be proved:

3) Every representation of G [including those of infinite dimension] is a direct sum of IRR's. [See Naimark.]

4) Let Γ be any faithful representation of G. Then all IRR's can be obtained by reducing

$$\Gamma,\ \Gamma^*,\ \Gamma \otimes \Gamma,\ \Gamma \otimes \Gamma^*,\ \ldots, \quad \Gamma \otimes \ldots \otimes \Gamma \otimes \Gamma^* \otimes \ldots \otimes \Gamma^*,\ \ldots$$

For noncompact groups the situation is much more complicated. Thus, for example, all faithful continuous unitary representations of a noncompact group are infinite dimensional. [For compact groups, all IRR's are finite dimensional.]

LECTURE 16

i) Lie algebras

Let us now recall the following covered in Lectures 14 and 15:

1) There exists a neighborhood U of e of G such that U generates G_0.

2) For compact G, if Γ is a faithful representation of G, we can get all IRR's by taking repeated direct products of Γ and Γ^* and reducing.

3) For compact G, the (left or right) regular representation contains all IRR's on reduction.

To find IRR's for compact groups:

a) 1) leads to the study of G in the "small" by using Lie algebra methods.

b) 2) leads to tensor methods.

c) 3) suggests the study of certain differential equations on the group [similar to those for the spin j rotation matrices].

We now take up Lie algebra and tensor methods in that order.

Definition. A Lie algebra L is a vector space on which a bilinear composition $[.,.]$ is defined. It fulfills the following properties:

1) Antisymmetry:

$$[x,\ y] = -[y,\ x].$$

2) Jacobi identity:

$$[x,\ [y,\ z]] + [y,\ [z,\ x]] + [z,\ [x,\ y]] = 0.$$

The dimension of L as a vector space is the dimension of the Lie algebra. We restrict L to be a real vector space for now.

Examples :

1) Consider all real antisymmetric 3×3 matrices with ordinary commutation of matrices as the Lie bracket. Call this L. Thus for x, $y \in L$, $[x, y] \equiv xy - yx \in L$. The Jacobi identity is clearly fulfilled. Also, dim L = 3. A basis for L is L_α, $\alpha = 1$, 2, 3 where

$$(L_\alpha)_{\beta\gamma} = -\varepsilon_{\alpha\beta\gamma}.$$

Note that

$$[L_\alpha, L_\beta] = \varepsilon_{\alpha\beta\gamma} L_\gamma.$$

(If $J_\alpha = iL_\alpha$, then $[J_\alpha, J_\beta] = i\varepsilon_{\alpha\beta\gamma} J_\gamma$. J_α's are the $J = 1$ angular momentum matrices in quantum mechanics.)

2) Consider all functions $\mathcal{F}(M)$ on a phase space M. Regard it as a vector space under pointwise addition. [$f_1 + f_2$ is defined by $(f_1 + f_2)(q,p) \equiv f_1(q,p) + f_2(q,p)$.] Define the bracket $[\alpha, \beta]$ for $\alpha, \beta \in \mathcal{F}(M)$ by

$$[\alpha, \beta](q, p) \equiv \sum_{i=1}^{N} \left[\frac{\partial\alpha}{\partial q_i} \frac{\partial\beta}{\partial p_i} - \frac{\partial\alpha}{\partial p_i} \frac{\partial\beta}{\partial q_i} \right] (q,p).$$

Then $\mathcal{F}(M)$ becomes an infinite-dimensional Lie algebra.

Definition. Let L be a Lie algebra with basis $\{e_\alpha\}$. Then since $[e_\alpha, e_\beta] \in L$, we can write

$$[e_\alpha, e_\beta] = C^\gamma_{\alpha\beta} e_\gamma,$$

where $C^\gamma_{\alpha\beta}$ are called the structure constants in the given basis. They fulfill

$$C^\gamma_{\alpha\beta} = -C^\gamma_{\beta\alpha}$$

and

$$C^\alpha_{\beta\mu} C^\mu_{\gamma\delta} + C^\alpha_{\gamma\mu} C^\mu_{\delta\beta} + C^\alpha_{\delta\mu} C^\mu_{\beta\gamma} = 0.$$

The latter follows from the Jacobi identity

$$[e_\alpha, [e_\beta, e\gamma]] + [e_\beta, [e_\gamma, e_\alpha]] + [e_\gamma, [e_\alpha, e_\beta]] = 0.$$

For the basis in the example of antisymmetric matrices, $e_\alpha = L_\alpha$, and since $[L_\alpha, L_\beta] = \varepsilon_{\alpha\beta\gamma} L_\gamma$, $C^\gamma{}_{\alpha\beta} = \varepsilon_{\alpha\beta\gamma}$.

Definition. A Lie algebra L' is homomorphic to a Lie algebra L if there exists a linear operator $T : L \to L'$ such that

$$[T(x), T(y)] = T([x, y]).$$

T is the homomorphism from L to L'.

Example:

Consider all traceless antihermitian 2×2 matrices with commutator bracket as Lie bracket. This Lie algebra L has basis $e_\alpha = -i\frac{\sigma_\alpha}{2}$, $\alpha = 1, 2, 3$ with

$$[e_\alpha, \ e_\beta] = \varepsilon_{\alpha\beta\gamma} \ e_\gamma.$$

Let $L' = \{\Sigma_\alpha \xi_\alpha L_\alpha\}$. Then

$$T\left(\sum_\alpha \xi_\alpha \ e_\alpha\right) = \sum_\alpha \xi_\alpha \ L_\alpha$$

is a homomorphism.

Definition. The homomorphism is an isomorphism if T is $1-1$. In the above example, T is an isomorphism.

Definition. Let L' be a Lie algebra of linear operators on a vector space V with commutator bracket as Lie bracket. Then it is a representation of a Lie algebra L if there is a homomorphism $L \to L'$. If the homormorphism is an isomorphism, the representation is faithful.

Examples:

1) The mapping

$$\sum_\alpha \xi_\alpha \left(-i\frac{\sigma_\alpha}{2}\right) \to \sum_\alpha \xi_\alpha L_\alpha$$

gives a faithful representation of $\{\sum_\alpha \xi_\alpha(-i\frac{\sigma_\alpha}{2})\}$ on a vector space of three dimensions.

2) Let $L = \{x\}$ be a Lie algebra. Define for each x, a linear operator ad x on L by

$$(\text{ad } x)y = [x, \ y]$$

so that each element of the Lie algebra is represented by a matrix acting on the vector space L. Then the Jacobi identity shows that

$$[\text{ad } x, \ \text{ad } y] = \text{ad}([x,y]).$$

It follows that ad : $x \to \text{ad } x$ gives a representation of L, the adjoint representation.

Moreover

$$(\text{ad } x)[y, \; z] = [(\text{ad } x)y, \; z] + [y, \; (\text{ad } x)z]$$

so that ad x behaves like a derivation for the commutator product.

Thus for $\{\sum_\alpha \xi_\alpha \left(-i\frac{\sigma_\alpha}{2}\right)\}$ with basis $e_\alpha = -i\frac{\sigma_\alpha}{2}$,

$$(\text{ad } e_\alpha)x = [e_\alpha, \; x].$$

Let us find the matrix of ad e_α in the basis $\{ e_\beta \}$:

$$(\text{ad } e_\alpha)e_\beta = [e_\alpha, \; e_\beta]$$
$$= \varepsilon_{\alpha\beta\gamma}e_\gamma$$
$$\equiv D(e_\alpha)_{\gamma\beta}e_\beta$$

or

$$D(e_\alpha)_{\gamma\beta} = \varepsilon_{\alpha\beta\gamma}$$
$$= -\varepsilon_{\alpha\gamma\beta}$$
$$= (L_\alpha)_{\gamma\beta}.$$

Thus $\{\sum_\alpha \xi_\alpha L_\alpha\}$ is the adjoint representation of $\{\sum_\alpha \xi_\alpha \left(-i\frac{\sigma_\alpha}{2}\right)\}$.

Quite generally as should be obvious, the matrix elements of ad e_α are given by the structure constants.

Remark:

The center $\mathcal{C}$ of a Lie algebra L is the set of all elements $c \in L$ which commute with every $l \in L$:

$$[c, \; l] = 0.$$

So the matrix ad $c = 0$.

It is easy to show that if $\mathcal{C}$ is trivial for L, $\mathcal{C} = \{0\}$, the representation ad is faithful.

The construction of ad shows that if $\mathcal{C}$ is trivial, $\mathcal{C} = \{0\}$, then any L can be represented faithfully by matrices.

ii) Equivalence of representations for Lie algebras

Let $L = \{\ell\}$ be a Lie algebra and $\{D^{(i)}(\ell)\}$ $(i=1, 2)$ be representations of L on vector spaces V_i. These representations are *equivalent* if there exists an invertible linear operator T from V_1 onto V_2 such that $TD^{(1)}(\ell) = D^{(2)}(\ell)T$, that is, $D^{(2)}(\ell) = T\, D^{(1)}(\ell)T^{-1}$.

Example:

Given a representation $L' = \{\ell'\}$ on V, if S is any nonsingular linear operator on V, then $L'' = \{S\ell'S^{-1}\}$ is equivalent to L'.

iii) Reducibility of representations for Lie algebras

A representation L' on V is reducible if there is a nontrivial subspace V_1 in V invariant under L', $L'V_1 \subset V_1$. [V_1 is nontrivial if V_1 is not $\{0\}$ or V.]

It is completely reducible if $V = V_1 \oplus V_2$ where V_i are nontrivial and $L'V_i \subset V_i$.

It is irreducible if there is no nontrivial subspace invariant under L'.

Schur's lemma

1) Let L', L'' be two IRR's of L on vector spaces V', V''. Let T be a linear operator from V' to V'' such that
$$T\ell'(x) = \ell''(x)T, \ \forall\, x \in L$$
where $\ell'(x)$, $\ell''(x)$ are representations of x in L', L''. Then either $T = 0$ or T is invertible and the representations are equivalent.

2) A linear operator which commutes with all the linear operators of an IRR is a multiple of the identity.

For proof, see Lecture 7.

Example:

Let $J_\alpha^{(j)}$ be the angular momentum matrices for spin j,
$$\left[J_\alpha^{(j)}, J_\beta^{(j)}\right] = i\, \varepsilon_{\alpha\beta\gamma} J_\gamma^{(j)}.$$
Then
$$\sum_\alpha \xi_\alpha \left(-i\, \frac{\sigma_\alpha}{2}\right) \to \sum_\alpha \xi_\alpha \left(-i\, J_\alpha^{(j)}\right)$$
defines a representation of $\left\{\sum_\alpha \xi_\alpha \left(-i\frac{\sigma_\alpha}{2}\right)\right\}$. It is irreducible. [For proof, see Lecture 25.]

iv) More on Lie groups

This part has some topological considerations which are not explained. They are included for completeness.

By similarity with the properties found for the general linear group $GL(n, \mathbb{R})$, we can introduce coordinate systems to parametrize the elements g of a continuous group G. [Thus G has a topology in which group operations are continuous.] A coordinate system $(x^1(g), x^2(g), \cdots, x^N(g)) \in \mathbb{R}^N$ is a continuous map with continuous inverse from a neighborhood of g, say U, to an open subset of $\mathbb{R}^N$ for a suitable N. The coordinate system is chosen such that the identity has its coordinates equal to zero. By considering elements of G in the intersection of two neighborhoods, it can be shown that N does not depend on the specific U. It is called the dimension of G. [This is a different notion of dimension from the one given in Lecture 1.]

We require that the composition map

$$\phi : G \times G \to G,$$

whenever it is restricted to any two neighborhoods

$$\phi : U \times V \to W,$$

is continuous in both the coordinates x, y of U, V, and that it is invertible whenever we fix one of the arguments. More specifically, we require that $\phi(x, y)$ be a continuous map for fixed x or y.

We obtain a Lie group if, in addition, ϕ is a real analytic map. [We will not go into the definition of real analyticity here. But see below.] By starting with a neighborhood U, and by taking the product of any pairs of elements in U, we construct U^2, from here U^3, and so on. The union of all these products is G_0. It is possible to show that G_0 is a group and does not depend on the starting neighborhood U (Pontrjagin). [See Lecture 14.]

For Lie groups, we consider neighborhoods U and V along with the composition map ϕ. We parametrize U with coordinates $(x^1, x^2, \cdots, x^N)$ and V with coordinates $(y^1, y^2, \cdots, y^N)$ as above. We set $\phi(x, y) = z$, with the assumptions

$$\phi(0, y) = y, \quad \phi(x, 0) = x.$$

Next consider the power series expansion of $\phi^j(x, y)$ around the identity of the group. [This expansion converges from the assumption of the real analyticity.] We find

$$\phi^j(x, y) = x^j + y^j + a^j_{lm} x^l y^m + r^j_{lmn} x^l x^m y^n + s^j_{lmn} x^l y^m y^n + \cdots.$$

Setting

$$q^j(x,y) = \phi^j(\phi(x,y), \phi(x^{-1}, y^{-1}))$$

the expansion gives

$$q^j(x,y) = c^j_{kl}x^k y^l + \cdots$$

where $c^j_{kl} = a^j_{kl} - a^j_{lk}$, and $c^j_{kl} = -c^j_{lk}$ are the "structure constants". It is given in terms of ϕ^j by

$$c^j_{kl} = \left(\frac{\partial^2 \phi^j}{\partial x^k \partial y^l} - \frac{\partial^2 \phi^j}{\partial x^l \partial y^k} \right) |_{x=y=0}.$$

A further property of these structure constants follows from the associative law for the group which can be formulated as

$$\phi(\phi(x,y), z) = \phi(x, \phi(y,z)).$$

Inserting the expansion of ϕ^j, in third order we find the quadratic relation

$$c^j_{kl}c^l_{mn} + c^j_{ml}c^l_{nk} + c^j_{nl}c^l_{km} = 0.$$

This relation will be used in Lecture 19. It is equivalent to the Jacobi identity.

LECTURE 17

i) Lie algebras as differential operators

In physics, we often represent the Lie algebra L of angular momentum matrices L_α by differential operators $\hat{L}_\alpha$:

$$\hat{L}_\alpha = (\vec{x} \wedge \vec{\nabla})_\alpha; = \epsilon_{\alpha\beta\gamma} x_\beta \partial_\gamma,$$

$$[\hat{L}_\alpha, \hat{L}_\beta] = \epsilon_{\alpha\beta\gamma} \hat{L}_\gamma.$$

This is an example of the "realization" or representation of a Lie algebra by differential operators acting on a vector space of functions.

In the above instance, we can work with functions f on $\mathbb{R}^3$ with

$$(\hat{L}_\alpha f)(x) = \epsilon_{\alpha\beta\gamma} x_\beta \frac{\partial f}{\partial x^\gamma}(x), \quad x = (x^1, x^2, x^3).$$

An important advantage of working with differential operators is the freedom to change coordinate systems without spoiling commutation relations. The change of coordinates can be nonlinear. In the above example, we can change coordinates on $\mathbb{R}^3$ and work for example with polar coordinates.

Such a differential realization exists whenever there is a representation T of L in terms of linear operators Lin (V, V) on a vector space V. If $\{L_\alpha\}$ is a basis for L and $y^1, y^2, \cdots, y^N$ are the coordinates of a vector y in a basis for V, then

$$(T(L_\alpha)y)^k = (A_\alpha)^k{}_j y^j.$$

Then on functions f on V, we have the following differential operator realization $\hat{T}(L_\alpha)$ of L_α:

$$(\hat{T}(L_\alpha)f)(y) = -\left[(A_\alpha)^k{}_j y^j \frac{\partial f}{\partial y^k}\right](y),$$

91

$$[\hat{T}(L_\alpha), \hat{T}(L_\beta)] = \hat{T}([L_\alpha, L_\beta]).$$

We can now freely change coordinates on V and still keep the representation $\hat{T}$ intact.

The realization $\hat{L}_\alpha$ of the angular momentum algebra is a particular example of this general case. For instance in Cartesian coordinates we have

$$\hat{L}_1 = -i(x_2 \frac{\partial}{\partial x_3} - x_3 \frac{\partial}{\partial x_2}),$$

$$\hat{L}_2 = -i(x_3 \frac{\partial}{\partial x_1} - x_1 \frac{\partial}{\partial x_3}),$$

$$\hat{L}_3 = -i(x_1 \frac{\partial}{\partial x_2} - x_2 \frac{\partial}{\partial x_1}).$$

In spherical polar coordinates, with

$$x_3 = r\cos\theta, \ x_2 = r\sin\theta\sin\phi, \ x_1 = r\sin\theta\cos\phi, \ r^2 = x_1^2 + x_2^2 + x_3^2,$$

we have

$$\hat{L}_1 = i(\sin\phi \frac{\partial}{\partial\theta} + \cos\phi\cot\theta \frac{\partial}{\partial\phi}),$$

$$\hat{L}_2 = -i(\cos\phi \frac{\partial}{\partial\theta} - \sin\phi\cot\theta \frac{\partial}{\partial\phi}),$$

$$\hat{L}_3 = -i\frac{\partial}{\partial\phi}.$$

The dual L^* of L consists of linear functions on L. Thus if $f \in L^*$, $\xi^\alpha e_\alpha \in L$ where $\{e_\alpha\}$ is a basis for L, then $f(\xi^\alpha e_\alpha) \in \mathbb{C}$. We can think of $\xi^\alpha e_\alpha$ as functions on L^* as well, that is as elements of L^{**}, the dual of the dual, by setting

$$\xi^\alpha e_\alpha(f) = f(\xi^\alpha e_\alpha).$$

Then this linear space $L^{**} = L$ of functions on L^* carries the Poisson bracket $(P.B.)$

$$\{\xi^\alpha e_\alpha, \eta^\beta e_\beta\} \, |_{P.B.} = \xi^\alpha \eta^\beta c^\gamma_{\alpha\beta} e_\gamma.$$

This idea can be made more explicit. In L^*, we can introduce the dual basis $\{f^\rho\}$ such that

$$e_\alpha(f^\rho) = \delta_\alpha{}^\rho.$$

So

$$e_\alpha(\sum \xi_\rho f^\rho) = \xi_\alpha.$$

Then

$$\{e_\alpha, e_\beta\}\left(\sum \xi_\rho f^\rho\right) = c_{\alpha\beta}^\gamma \xi_\gamma.$$

Now a point $\sum \xi_\rho f^\rho$ of L^* can be described by $\xi = (\xi_1, \xi_2, \cdots, \xi_N)$. Then for $\alpha \in L$, we can write $\alpha(\sum \xi_\rho f^\rho)$ as just $\alpha(\xi)$ and the above $P.B.$ as

$$\{e_\alpha, e_\beta\}(\xi) = \xi_\gamma c_{\rho\sigma}^\gamma \frac{\partial e_\alpha(\xi)}{\partial \xi_\rho} \frac{\partial e_\beta(\xi)}{\partial \xi_\sigma}.$$

Here e_α's and α's are in L, and hence linear functions on L^*. But for generic nonlinear functions α, β on L^* as well, this formula gives a $P.B.$:

$$\{\alpha, \beta\}(\xi) = \xi_\gamma c_{\rho\sigma}^\gamma \frac{\partial \alpha(\xi)}{\partial \xi_\rho} \frac{\partial \beta(\xi)}{\partial \xi_\sigma}.$$

In this formalism, we can also freely change ξ by coordinate transformations as we do in classical mechanics.

In $\mathbb{R}^3$ the Lie algebra of the rotation group may be realized in terms of Poisson brackets as

$$\{x, y\} = z, \ \{y, z\} = x, \ \{z, x\} = y.$$

This may be extended to general smooth functions by setting

$$\{f, g\} = \frac{\partial f}{\partial x}\{x, y\}\frac{\partial g}{\partial y} - \frac{\partial f}{\partial y}\{x, y\}\frac{\partial g}{\partial x} + \frac{\partial f}{\partial y}\{y, z\}\frac{\partial g}{\partial z}$$
$$- \frac{\partial f}{\partial z}\{y, z\}\frac{\partial g}{\partial y} + \frac{\partial f}{\partial z}\{z, x\}\frac{\partial g}{\partial x} - \frac{\partial f}{\partial x}\{z, x\}\frac{\partial g}{\partial z}.$$

In more compact notation, by setting $\mathrm{grad} f = (\frac{\partial f}{\partial x}, \frac{\partial f}{\partial y}, \frac{\partial f}{\partial z})$, we have

$$\{f, g\} = (\mathrm{grad} f) \begin{bmatrix} 0 & z & -y \\ -z & 0 & x \\ y & -x & 0 \end{bmatrix} (\mathrm{grad} g)^t$$

$$-(\mathrm{grad} g) \begin{bmatrix} 0 & z & -y \\ -z & 0 & x \\ y & -x & 0 \end{bmatrix} (\mathrm{grad} f)^t.$$

The matrix $P = \begin{bmatrix} 0 & z & -y \\ -z & 0 & x \\ y & -x & 0 \end{bmatrix}$ is called the Poisson structure matrix. If we introduce spherical polar coordinates, we find the Poisson bracket in these coordinates as given by

$$\{r, \theta\} = 0, \ \{r, \phi\} = 0, \ \{\theta, \phi\} = \frac{1}{r \sin \theta}.$$

ii) More on Lie algebras as differential operators

By using the matrix of the Poisson structure, it is possible to represent all three-dimensional real Lie algebras. Thus for

$$P = \begin{bmatrix} 0 & cz & -by \\ -cz & 0 & ax \\ by & -ax & 0 \end{bmatrix},$$

the corresponding bracket is

$$\{f, g\} = (\mathrm{grad} f)(P)(\mathrm{grad} g)^t - (\mathrm{grad} g)(P)(\mathrm{grad} f)^t.$$

When a, b, c are given real values, say 0, ± 1, we get the various 3-dimensional Lie algebras (of type A in Bianchi's classification (see Bibliography for references)). When

$$P = \begin{bmatrix} 0 & 0 & -by - mx \\ 0 & 0 & ax - my \\ mx + by & my - ax & 0 \end{bmatrix},$$

in a similar way, we get all Lie algebras of type B.

What we want to stress is that the appropriate classification of the Poisson matrix structures allows us to enumerate all possible 3-dimensional real Lie algebras. This procedure extends to any dimension, even though the method to find P becomes more involved because it must satisfy a quadratic relation coming from the Jacobi identity.

By writing the Poisson bracket in the form

$$\{f, g\} = (\mathrm{grad} f)(P)(\mathrm{grad} g)^t - (\mathrm{grad} g)(P)(\mathrm{grad} f)^t,$$

we may consider, in addition to Poisson brackets defined by matrix structures linear in the coordinates, Poisson brackets associated with matrices P with generic polynomial entries.

A particular example of a quadratic bracket is provided, in 3 dimensions, by the following matrix:

$$P = \begin{bmatrix} 0 & 3cz^2 & -3by^2 - mx^2 - 2xy \\ -3cz^2 & 0 & 3ax^2 + 2mxy + my^2 \\ 3by^2 + mx^2 + 2xy & -3ax^2 - 2mxy - my^2 & 0 \end{bmatrix}.$$

We conclude this remark by saying that realizing Lie algebras in terms of Poisson brackets may have interesting generalizations.

LECTURE 18

i) On canonical realizations of Lie algebras

Very often both in classical and quantum mechanics we consider canonical coordinates and canonical commutation relations respectively. It is therefore very convenient to realize Lie algebras in terms of functions of canonical coordinates for classical mechanics or operator functions of creation and annihilation operators in quantum theory.

To be specific, we may consider canonical coordinates (q_1, p_1, q_2, p_2) on $\mathbb{R}^4$ and search for functions $f_{x,y,z}$ where

$$\{f_x, f_y\} = f_z, \ \{f_y, f_z\} = f_x, \{f_z, f_x\} = f_y,$$

where in the computation we use $\{q_a, p_b\} = \delta_{ab}$, $\{p_a, p_b\} = 0$, $\{q_a, q_b\} = 0$.

This formulation is a classical instance of the so-called Jordan-Schwinger realization of 3-dimensional Lie algebras in terms of creation and annihilation operators.

It is not difficult to show that the following quadratic expression provides a realization of the angular momentum algebra for the classical and the quantum situation, respectively:

Classical:

$$f_x = \frac{1}{2}(q_1 q_2 + p_1 p_2), \ f_y = \frac{1}{2}(q_1 p_2 - q_2 p_1), \ f_z = \frac{1}{4}(q_1^2 + p_1^2 - q_2^2 - p_2^2).$$

Casimir function (with zero Poisson bracket with $f_{x,y,z}$):

$$\mathcal{C} = p_1^2 + q_1^2 + p_2^2 + q_2^2.$$

Quantum:

$$\hat{f}_x = \frac{1}{2}(\hat{q}_1 \hat{q}_2 + \hat{p}_1 \hat{p}_2), \ \hat{f}_y = \frac{1}{2}(\hat{q}_1 \hat{p}_2 - \hat{q}_2 \hat{p}_1), \ \hat{f}_z = \frac{1}{4}(\hat{q}_1^2 + \hat{p}_1^2 - \hat{q}_2^2 - \hat{p}_2^2).$$

Casimir operator:

$$\hat{C} = \hat{p}_1^2 + \hat{q}_1^2 + \hat{p}_2^2 + \hat{q}_2^2.$$

The realization in terms of Poisson brackets allows us to consider not only the commutation relations among linear function elements of the Lie algebra, but also among polynomial functions.

In general the corresponding Lie algebra will be infinite dimensional. But when we consider the Poisson bracket among quadratic functions of (q_1, p_1, q_2, p_2), we get as a result again a quadratic function. Therefore the space of quadratic functions in the variables (q_1, p_1, q_2, p_2) is a Lie algebra (indeed it is the Lie algebra of the symplectic group in two dimensions).

However if we consider the Poisson bracket of two cubic functions, we get a quartic function, and so on. Therefore the Lie algebra becomes infinite dimensional.

A similar situation prevails when we consider polynomial expressions of $(a_1, a_1^\dagger, a_2, a_2^\dagger)$, i.e. annihilation and creation operators.

The extension of the "Lie algebra brackets" to polynomial functions in general allows for the definition of Casimir operators as those which commute with any element of the Lie algebra. From the point of view of the representation theory it means that the eigenvalues of Casimir operators can provide labels of irreducible representations.

ii) On harmonic polynomials

As an application of realizations of Lie algebras in terms of differential operators, we now consider spherical harmonics.

Consider the space of polynomials $\mathbb{C}[x_1, x_2, x_3]$ on $\mathbb{R}^3$ with complex coefficients. This infinite dimensional vector space may be given the structure of a Hilbert space if we consider the inner product defined by

$$\langle \mathcal{P}_1 | \mathcal{P}_2 \rangle = \int_{\mathbb{R}^3} \mathcal{P}_1^* \mathcal{P}_2 \exp\left(-(x_1^2 + x_2^2 + x_3^2)\right) dx_1 dx_2 dx_3.$$

This space decomposes into orthogonal subspaces of homogeneous polynomials.

$$\mathcal{P} \equiv \oplus_k \mathcal{P}_k.$$

Each differential operator $R_j = \epsilon_{jkl} x^k \frac{\partial}{\partial x^l}$ preserves the degree of homogeneous polynomials. Therefore their corresponding matrix in $\mathcal{P}$ will have a block-diagonal form.

For instance quadratic polynomials, generated by x_1^2, x_2^2, x_3^2, x_1x_2, x_2x_3, x_3x_1 will be mapped into themselves and the corresponding linear transformation will act preserving the one-dimensional subspace generated by $x_1^2 + x_2^2 + x_3^2$ and the 5-dimensional complementary subspace generated by $x_1^2 - x_2^2$, $x_2^2 - x_3^2$, x_1x_2, x_2x_3, x_3x_1. The corresponding 6×6 matrix of each infinitesimal generator of rotations will split into a 5×5 matrix and a 1×1 one.

Similarly for cubic polynomials, generated by x_1^3, x_2^3, x_3^3, $x_1^2x_2$, $x_1x_2^2$, $x_2^2x_3$, $x_2x_3^2$, $x_3^2x_1$, $x_3x_1^2$, $x_1x_2x_3$, the space is 10-dimensional and splits into two invariant subspaces of dimensions 3 and 7 respectively. They are generated by x_1r^2, x_2r^2, x_3r^2, $r^2 = x_1^2 + x_2^2 + x_3^2$, and a complementary one, orthogonal to it for the scalar product above.

In fact, each subspace $\mathcal{P}_{2k}$ of homogeneous polynomials of even degree $2k$ and each subspace $\mathcal{P}_{2k+1}$ of homogeneous polynomials of odd degree $2k + 1$ will decompose into orthogonal invariant subspaces,

$$\mathcal{P}_{2k+1} = H_1 r^{2k} \oplus H_3 r^{2(k-1)} \oplus \cdots \oplus H_{2(k-1)+1} r^2 \oplus H_{2k+1},$$

$$\mathcal{P}_{2k} = H_0 r^{2k} \oplus H_2 r^{2(k-1)} \oplus \cdots \oplus H_{2(k-1)} r^2 \oplus H_{2k}.$$

Here $\mathcal{P}_k = H_k r^k$ and we restrict r to be > 0.

Due to the invariance of any polynomial of r^2 under rotations, each H_k is invariant under rotations.

By using the expression of the Laplacian $\triangle f$ as div grad$f$, we find

$$\triangle(\vec{r} \cdot \vec{a})^l = \text{div grad}(\vec{r} \cdot \vec{a})^l = \text{div}(l\vec{a}(\vec{r} \cdot \vec{a})^{l-1}) = l(l-1)\vec{a} \cdot \vec{a}(\vec{r} \cdot \vec{a})^{l-2}.$$

In particular, if the vector $\vec{a}$ is chosen with complex coefficients, say $\vec{a} = (i(t^2 + 1), (t^2 - 1), 2t)$ such that $\vec{a} \cdot \vec{a} = -(t^2 + 1)^2 + (t^2 - 1)^2 + (2t)^2 = 0$, we obtain

$$\left(i(t^2 + 1)x + (t^2 - 1)y + 2tz\right)^l = \left(t^2(y + ix) - (y - ix) + 2tz\right)^l$$

$$= t^l \sum_{m=-l}^{l} t^m H_{lm}(\vec{r})$$

and H_{lm} will be the generators of homogeneous polynomials of degree l which generate the spaces of H_l of our previous decomposition. They are $2l+1$ in number and they are in the kernel of $\triangle$, that is, they are annihilated by $\triangle$. They are called harmonic polynomials. When each of these harmonic polynomials is restricted to the unit sphere in $\mathbb{R}^3$, we obtain the spherical harmonics Y_{lm}. By using the familiar expression of the vector $\vec{r}$ in radial and polar coordinates, for $r = 1$, we have

$$\left(i(t^2 + 1)\sin\theta\cos\phi + (t^2 - 1)\sin\theta\sin\phi + 2t\cos\theta\right)^l = t^l \sum_{m=-l}^{l} t^m Y_{lm}(\theta, \phi).$$

Clearly this presentation can be generalized to any dimension and allows for the introduction of "hyperspherical" functions.

LECTURE 19

i) Relation between Lie groups and Lie algebras

Below, we first discuss how to associate a Lie algebra L to every Lie group G_0. Then we discuss the reconstruction of a) G_0 from L and b) the representations of G_0 from those of L. While few of the results are proved, they are illustrated by simple examples.

We assume that G_0 is a connected Lie group.

1) One-parameter subgroups: Suppose $g[\xi(t)] \equiv g(t)$, $t \in \mathbb{R}^1$ is a curve in G_0 such that $g(0) = e$ and $g(s)\,g(t) = g(s+t)$. Then $\{g(t)\}$ is a one-parameter (Abelian) subgroup of G_0. [Here $\xi = (\xi^1,\ \xi^2,\ \dots,\ \xi^n)$ denotes the coordinates for G_0.]

Example:

For $G_0 = SO(3)$,
$$\left\{ e^{it(\hat{n}\cdot\vec{J})} \big| t \in \mathbb{R}^1,\ \hat{n} = \text{ fixed unit vector} \right\}$$
is a one-parameter subgroup isomorphic to $SO(2)$ or $U(1)$.

Note that such a subgroup is generated by
$$\{g(t) |\, |t| < \delta,\ \text{any } \delta > 0\}.$$

2) Recall the group germ U of Lecture 14. The theorem we need is: we can choose U such that every element in U lies on a one-parameter subgroup (cf. Talman). Thus if we take (arbitrarily small) neighborhoods of the origin of all one-parameter subgroups, they generate G_0.

Example:

$G_0 = $ SO(3). Every g lies on a one-parameter subgroup. For by Euler's theorem, $g = e^{i\theta \hat{n} \cdot \vec{J}}$ for some θ and $\hat{n}$. Therefore $g \in \left\{ e^{it\hat{n} \cdot \vec{J}} \right\}$.

For all compact groups, every g lies on a one-parameter subgroup. This is not necessarily true for noncompact groups (cf. Talman).

3) Let $\{g(t) \equiv g[\xi(t)]\}$ be a one-parameter subgroup. Then it is uniquely fixed by its tangent vector at the origin, namely

$$\dot{\xi}(0) = (\dot{\xi}^1(0), \ \dot{\xi}^2(0), \ \dots, \ \dot{\xi}^n(0)).$$

Example:

$G_0 = SO(3)$. Let $g(t) = e^{it\hat{n} \cdot \vec{J}}$. The tangent vector at the origin to $t\hat{n}$ is $\hat{n}$. Knowing $\hat{n}$, we can reconstruct the one-parameter subgroup $\{e^{it\hat{n} \cdot \vec{J}}\}$.

4) Let $\phi = (\phi^1, \ \dots, \phi^n)$ define the group composition law:

$$g(\xi)g(\eta) = g(\phi(\xi, \eta))$$

Define

$$C^\alpha_{\beta\gamma} = \left(\frac{\partial^2 \phi^\alpha(\xi, \eta)}{\partial \xi^\beta \partial \eta^\gamma} - \frac{\partial^2 \phi^\alpha(\xi, \eta)}{\partial \xi^\gamma \partial \eta^\beta} \right) \Big|_{\xi=\eta=0}.$$

Then $C^\alpha_{\beta\gamma} = -C^\alpha_{\gamma\beta}$. Also, associativity of the group composition implies the Jacobi identity

$$C^\alpha_{\beta\mu} C^\mu_{\gamma\delta} + C^\alpha_{\gamma\mu} C^\mu_{\delta\beta} + C^\alpha_{\delta\mu} C^\mu_{\beta\gamma} = 0.$$

[See iv), Lecture 16.]

Now, for every vector $\dot{\xi}(0)$, there exists a one-parameter subgroup for which it is the tangent vector at the origin.

Example:

$G_0 = SO(3)$. Consider the one-parameter subgroup $\{ e^{it\xi \hat{n} \cdot \vec{J}} \mid \xi$ fixed, real $\}$. Its tangent is $\xi \hat{n}$ which can be any real three-vector.

The set of tangent vectors at the origin thus forms an n-dimensional real vector space, $n = \dim G_0$. Now make it into a real Lie algebra L by defining the vector $[\xi, \ \eta]$ for every pair of vectors ξ and η by

$$[\xi, \ \eta]^\alpha = C^\alpha_{\beta\gamma} \xi_\beta \eta_\gamma.$$

[This bracket is antisymmetric and fulfills the Jacobi identity because of the properties of $C^\alpha_{\beta\gamma}$ stated above.]

5) For matrix groups, every one-parameter subgroup is of the form

$$\{e^{tU} \mid U \ \text{fixed}\}.$$

The allowed U's form an n-dimensional vector space.

Example:

$G_0 = SO(3)$. A basis for the U's is

$$L_\alpha = -iJ_\alpha, \quad (J_\alpha)_{ij} = -i\varepsilon_{\alpha ij}.$$

6) The allowed U's can be made into a Lie algebra L' with Lie bracket given by ordinary commutation. L' is isomorphic to L.

Example:

$G_0 = SO(3)$. If $U = \xi_\alpha(-iJ_\alpha)$ and $V = \eta_\beta(-iJ_\beta)$, then

$$[U,\, V] = \xi_\alpha \eta_\beta[-iJ_\alpha,\, -iJ_\beta]$$
$$= \xi_\alpha \eta_\beta \varepsilon_{\alpha\beta\gamma}(-iJ_\gamma).$$

7) Given an n-dimensional real Lie algebra, there exists a (local) Lie group of which it is the Lie algebra in the sense of the foregoing. That is, given $C^\alpha_{\beta\gamma}$ [real, antisymmetric, and fulfilling Jacobi identity], there exists $\phi = (\phi^1,\, \ldots,\, \phi^n)$ such that if $x,\, y \in \mathbb{R}^n$, we can define a composition $x \circ y$ where

$$(x \circ y)^\alpha = \phi^\alpha(x, y).$$

Under this composition, the x's fulfill the group properties for sufficiently small $|x|$, $|y|$ (where $|x| = (\sum_i x_i{}^2)^{1/2}$, $|y| = (\sum_i y_i{}^2)^{1/2}$).

8) Given two connected Lie groups G_0, H_0 with a homomorphism τ: $G_0 \rightarrow H_0$, there is a homomorphism from the Lie algebra of G_0 to that of H_0. Also if τ has a discrete kernel [i.e. dim $G_0 = $ dim H_0], the Lie algebra homomorphism is an isomorphism.

Example:

$G_0 = SU(2)$, $H_0 = SO(3)$. The Lie algebra of G_0 is spanned by $-i\frac{\sigma_\alpha}{2}$, that of H_0 is spanned by $-iJ_\alpha$ and

$$\sum_\alpha \xi_\alpha(-i\frac{\sigma_\alpha}{2}) \rightarrow \sum_\alpha \xi_\alpha(-iJ_\alpha)$$

is an isomorphism.

If τ has a discrete kernel K, we say that G_0 is a covering group of H_0, covering it $|K|$ times. [Here $|K|$ is the number of elements in K.]

Example:

We say that $SU(2)$ is the covering group of $SO(3)$, covering it twice.

9) For every connected Lie group G_0, there exists a unique (upto isomorphism) connected and simply connected Lie group $\bar{G}_0$ of the same dimension as G_0 such that there is a homomorphism $\tau\colon \bar{G}_0 \ \to \ G_0$ with discrete kernel K. The number of times $\bar{G}_0$ covers $G_0(\equiv | \ K \ |)$ is equal to the connectivity of G_0. If any connected Lie group "covers" $\bar{G}_0$ with a discrete kernel K, it is isomorphic to $\bar{G}_0$ (i.e., $K = \{e\}$). [We shall explain the meaning of "connectivity" in the next lecture.]

The group $\bar{G}_0$ is the *universal covering group* (UCG) of G_0.

10) All connected Lie groups with isomorphic Lie algebras have the same universal covering group.

11) The converse of 8) is not always true. Thus if G_i are connected Lie groups with Lie algebras L_i, the existence of the homomorphism $L_1 \to L_2$ does not always imply the existence of the homomorphism $G_1 \to G_2$. (See Lecture 23.) However, we can say the following: Let $\bar{G}_0$ be the UCG of G_1 or G_2 with the associated Lie algebra $\bar{L}$. Then there are homomorphisms $\bar{G}_0 \to G_1$, $\bar{G}_0 \to G_2$. By 8), there are isomorphisms $\bar{L} \to L_1, \bar{L} \to L_2$ since $\dim G_i = \dim \bar{G}_0$.

12) If the Lie groups G_i have Lie algebras L_i, then the Lie algebra of $G_1 \times G_2 \times \ \ldots$ is $L_1 \oplus L_2 \oplus \ldots$. Here if the subspace L_i of the direct sum $\oplus_i L_i$ is spanned by $\{e_\nu^{(i)}\}$, $[e_\nu^{(i)}, \ e_\lambda^{(j)}] = 0$ if $i \neq j$ while $[e_\nu^{(i)}, \ e_\lambda^{(j)}]$ are those appropriate for the Lie algebra of G_i.

The result 12) is readily proved from the structure of ϕ for direct products of groups and the definition of $C_{\beta\gamma}^{\alpha}$ in terms of ϕ.

LECTURE 20

i) More on one-parameter subgroups 1

We saw that the fundamental idea which lets us associate Lie algebras to Lie groups relies on the notion of one-parameter subgroups. We now repeat some considerations of the previous lecture in a slightly different language and give additional examples.

A one-parameter subgroup is given by a map $\xi : \mathbb{R} \times G_0$ to G_0. We write

$$\xi : (t, g) \to g(t).$$

The map fulfills the additional properties
1) $\xi : (0, g) \to e$ or $g(0) = e$.
2) $g(s)g(t) = g(s + t)$.
3) Now on any smooth function $f \in \mathcal{F}(G)$ on G, we have the left-(say) regular representation of G. Set

$$(X_g f)(h) = \frac{d}{ds} f(g(s)^{-1}h) \mid_{s=0} .$$

Then X_g is a homogeneous first-order differential operator.

Thus to every one-parameter subgroup, we have a homogeneous first-order differential operator on $\mathcal{F}(G)$. It is the tangent to the identity of the one-parameter subgroup. Conversely by integrating ("exponential map"), we can recover the one-parameter subgroup. Its action on $\mathcal{F}(G)$ is given by

$$e^{tX_g} f = \sum \frac{t^n}{n!} X_g^n f.$$

One can check that

$$(e^{tX_g} f)(h) = f(g(t)^{-1}h).$$

We can change the parametrization t so that $t = 0$ corresponds to any element g_0 on the given one-parameter subgroup. For this, it is enough to consider $\{\phi(t) = g(t)g_0\}$, $g(0) = e$. The original tangent is then

$$\frac{d\phi(t)}{dt}\phi(t)^{-1}\,|_{t=0} = X_g.$$

It is sometimes called the logarithmic derivative of ϕ.

Example:

An example of the exponential map acting on functions is

$$(e^{a\frac{d}{dx}}f)(x_0) = f(x_0 + a) = f(x_0) + af'(x_0) + \frac{a^2}{2!}f''(x_0) + \cdots.$$

ii) More on one-parameter subgroups 2

We saw that with any one-parameter subgroup, we can associate a homogeneous first-order differential operator. It is called an infinitesimal generator of the one-parameter subgroup. We can recover the latter from the former as well.

As an example, consider the transformation group

$$\{A(\alpha) = \begin{pmatrix} \cosh\alpha & \sinh\alpha \\ \sinh\alpha & \cosh\alpha \end{pmatrix} \mid \alpha \in \mathbb{R}\}$$

acting on $\mathbb{R}^2$. Under this transformation, the coordinate $x = \begin{pmatrix} x_1 \\ x_2 \end{pmatrix}$ becomes

$$x(\alpha) \equiv \begin{pmatrix} x_1(\alpha) \\ x_2(\alpha) \end{pmatrix} = \begin{pmatrix} \cosh\alpha & \sinh\alpha \\ \sinh\alpha & \cosh\alpha \end{pmatrix}\begin{pmatrix} x_1 \\ x_2 \end{pmatrix}.$$

This $x(\alpha)$ can be obtained by integrating the ordinary first order differential equation

$$\frac{d}{d\alpha}\begin{pmatrix} x_1(\alpha) \\ x_2(\alpha) \end{pmatrix} = \begin{pmatrix} 0 & 1 \\ 1 & 0 \end{pmatrix}\begin{pmatrix} x_1(\alpha) \\ x_2(\alpha) \end{pmatrix}.$$

The corresponding equation on the Lie algebra has the form

$$\frac{dA(\alpha)}{d\alpha}A^{-1}(\alpha) = \begin{pmatrix} 0 & 1 \\ 1 & 0 \end{pmatrix}$$

whose solution is

$$A(\alpha) = \exp\left(\alpha \begin{pmatrix} 0 & 1 \\ 1 & 0 \end{pmatrix}\right)$$

for the initial condition $A(0) = \mathbf{1}$. It satisfies all the properties we have required for a one-parameter subgroup.

As mentioned above, the expression

$$\left(\frac{d}{d\alpha} A(\alpha)\right) A^{-1}(\alpha)$$

is called the logarithmic derivative of $A(\alpha)$. It can be defined for any connected Lie group and provides us with the map from a Lie group to its Lie algebra.

The exponential map, going from the infinitesimal generator to the one-parameter subgroup, provides us the inverse map: from Lie algebras to Lie groups.

LECTURE 21

i) Connectivity

Let us now explain the meaning of connectivity.

Let G_0 be a connected Lie group. Consider all closed curves in G_0 starting and ending at some fixed point, say e. Then they can be decomposed into sets (equivalence classes) C_i ($i \in$ some index set) such that:

1) Any two closed curves in C_i can be continuously deformed into each other.

2) No curve in C_i can be deformed to a curve in C_j if $i \neq j$.

The number of such classes is called the connectivity of G_0.

G_0 is simply connected if there is only one such class. Then any closed curve can be deformed to a point.

Examples:

1) $G_0 = U(1)$. $U(1) = \{e^{i\phi} | \ \phi \text{ real}\}$ is the same as the circle S^1 topologically. The fixed point is taken to be 1. Let $\Delta\phi$ denote the change of ϕ as a closed curve (from 1 to 1) is traversed on $U(1)$. Then for closed curves of class C_n, $\Delta\phi = 2\pi n (n = 0, \ \pm 1, \ \pm 2, \ \ldots)$. It is obvious that curves of class C_n can be deformed into each other and that they cannot be deformed into curves of class C_m if $m \neq n$. Thus $U(1)$ is infinitely connected.

The universal covering group of $U(1)$ is the translation group $T_1 = \{a | a \in \mathbb{R}^1\}$. The homomorphism $T_1 \to U(1)$ is $a \to e^{ia}$. It has the discrete kernel

$$\{a = 2\pi n | \ n = 0, \ \pm 1, \ \pm 2, \ \ldots\}$$

whose cardinality is the same as the connectivity of $U(1)$. Any closed curve in T_1 can be shrunk to a point, so that T_1 is simply connected.

2) $G_0 = SO(3)$. Recall the geometry of $SO(3)$: It can be represented

as the ball $\{x \in \mathbb{R}^3 | [\sum_i (x^i)^2]^{1/2} \leq \pi\}$, diametrically opposite points on the surface of the ball being identified. For a point P of the ball, the direction of the line from the origin to P is the axis of rotation, its length is the angle of rotation. [All rotations are performed in the clockwise direction.] The rotations by $\pm\pi$ around an axis are equal leading to the identification stated above.

Let us examine closed curves in $SO(3)$;

(0) 0 *jump curves.*

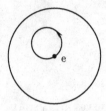

This curve can be deformed to e.

(1) 1 *jump curves.*

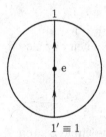

This curve cannot be deformed to (0).

(2) 2 *jump curves.*

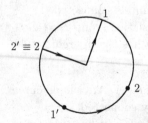

This can be deformed to (0): Deform 2 to $1'(2'$ to 1). The curve is now

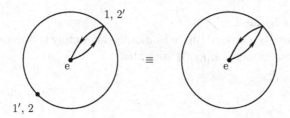

which can be deformed to (0). Thus curves with an even number of jumps can be deformed to (0). Curves with an odd number of jumps can be deformed to (1). Consequently $SO(3)$ is doubly connected.

$SU(2)$ is connected [use the Euler decomposition

$$e^{i\alpha\tau_3} \; e^{i\beta\tau_2} \; e^{i\gamma\tau_3}$$

and shrink each factor to the identity by decreasing the angle to zero], dim $SU(2) = \dim SO(3)$ and there exists a homomorphism $SU(2) \to SO(3)$ with a kernel $\mathbb{Z}_2(\equiv S_2)[= $ connectivity of $SO(3)]$. This implies that $SU(2)$ is simply connected. It is the universal covering group of $SO(3)$.

ii) Summary of relations between Lie groups and Lie algebras

Notation
$A \to B \equiv$ Homomorphism from A to B.
$A \rightleftharpoons B \equiv$ Isomorphism between A and B.
$G \rightsquigarrow L \equiv L$ is the Lie algebra of G.
$K \quad \equiv$ Kernel of the map.

All groups are connected Lie groups of the *same* dimension (for simplicity). $\bar{G}_0$ and $\bar{L}$ are the universal covering group and its Lie algebra.

1) $G_1 \to G_2 \to G_3 \cdots$

$$\begin{array}{ccc} \rightsquigarrow & \rightsquigarrow & \rightsquigarrow \\ \downarrow & \downarrow & \downarrow \\ L_1 & L_2 & L_3 \end{array} \Rightarrow L_1 \rightleftharpoons L_2 \rightleftharpoons L_3.$$

2)
$$G_1 \quad G_2 \quad G_3 \cdots$$

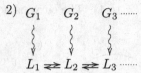

$$L_1 \rightleftarrows L_2 \rightleftarrows L_3 \cdots$$

$\Rightarrow$ there exists a unique (up to isomorphism) simply connected Lie group $\bar{G}_0$ of the same dimension as G_i such that

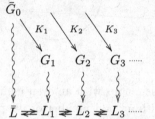

$$\bar{G}_0$$
$$\quad K_1 \quad K_2 \quad K_3$$
$$G_1 \quad G_2 \quad G_3 \cdots$$
$$\bar{L} \rightleftarrows L_1 \rightleftarrows L_2 \rightleftarrows L_3 \cdots$$

K_i are discrete sets of dimension equal to the connectivity of G_i.

Examples:

1) $G_0 \equiv SU(2) \longrightarrow SO(3)$

$$\bar{L} \equiv L_{SU(2)} \Longleftrightarrow L_{SO(3)}.$$

2) $\bar{G}_0 \quad \equiv \quad SU(2) \quad \times \quad SU(2) > L_{SU(2) \times SU(2)} = \bar{L}$

$\quad \downarrow K_1 \qquad\qquad \downarrow K_2 \quad | K_3 \quad | K_4$

$SU(2) \times SO(3) \leadsto\!\longrightarrow L_{SU(2) \times SO(3)}$

$SO(3) \times SU(2) \leadsto\!\longrightarrow L_{SO(3) \times SU(2)}$

$SO(4) \leadsto\!\longrightarrow L_{SO(4)}$

$\quad K_5$

$SO(3) \times SO(3) > L_{SO(3) \times SO(3)}$

and

$$SU(2) \times SO(3) \xrightleftharpoons{\tau} SO(3) \times SU(2)$$
$$\quad K_6 \qquad K_7$$
$$SO(3) \times SO(3).$$

The group $SU(2) \times SU(2)$ has the invariant subgroup [center]

$$\mathcal{C} = \{e_{++} = (1, \ 1), \ e_{+-} = (1, \ -1), \ e_{-+} = (-1, \ 1), \ e_{--} = (-1, \ -1)\}.$$

The K_i are given by

$$K_1 = \{e_{++}, e_{+-}\}, \qquad K_2 = \{e_{++}, e_{-+}\},$$
$$K_3 = \{e_{++}, e_{--}\}, \qquad K_4 = \mathcal{C},$$
$$K_5 = \{(1, -1)\}, \qquad K_6 = \{(1, \ 1), \ (-1, \ 1)\},$$
$$K_7 = \{(1, 1), \ (1, -1)\}.$$

In K_5, 1 and -1 are 4×4 matrices. The isomorphism τ is given by

$$SU(2) \times SO(3) \ni (g, \ R) \overset{\tau}{\to} (R, \ g) \in SO(3) \times SU(2).$$

The homomorphism $SU(2) \times SU(2) \to SO(4)$ is given by

$$SU(2) \times SU(2) \ni (g_1, g_2) \to g_1 \otimes g_2 \in SO(4)$$

where $g_1 \otimes g_2$ denotes the Kronecker product of the matrices g_1 and g_2. The matrices $\{g_1 \otimes g_2\}$ can be made real and orthogonal by a change of basis.

A little thought shows that there is no homomorphism connecting $SU(2) \times SU(2)/K_3 \equiv SO(4)$ to either $SU(2) \times SU(2)/K_1 \equiv SU(2) \times SO(3)$ or $SU(2) \times SU(2)/K_2 \equiv SO(3) \times SU(2)$. This illustrates point 2) above.

There are slight abuses of notation in the preceding discussion. Thus, for example, in the equation defining K_6, e_{-+} is the element $(-1, 1)$ of $SU(2) \times SO(3)$ while in the equation defining $\mathcal{C}$, e_{-+} is the element $(-1, 1)$ of $SU(2) \times SU(2)$. These minor imprecisions however should not cause confusion.

LECTURE 22

i) Representation theory

Let $L = \{\ell\}$ be the Lie algebra of a connected Lie group G_0 and let $\gamma = \{f(\ell)\}$ be a representation of L. Then the exponentials $e^{tf(\ell)}$ generate a connected Lie group Γ with γ as its Lie algebra. The existence of the homomorphism $\ell \to f(\ell)$ then implies (from previous statements) that Γ is a representation of the UCG $\bar{G}_0$. Thus finding representations of the Lie algebra is equivalent to finding representations of the UCG. [The exponential $e^{tf(\ell)}$ is well-defined by its power series if γ is finite dimensional. If it is infinite dimensional, $e^{tf(\ell)}$ may not exist and the preceding statements have to be qualified.]

Example:

$G_0 = SO(3)$. $L_{SO(3)}$ has basis L_1, L_2, L_3 with

$$[L_i,\, L_j] = \varepsilon_{ijk}\, L_k.$$

The IRR's $\gamma^{(j)}$ of $L_{SO(3)}$ are i times spin j angular momentum matrices ($j = 0, 1/2, 1, 3/2, \ldots$). Let $\Gamma^{(j)}$ be the group associated with $\gamma^{(j)}$. Then $\Gamma^{(j)}$ are representations of $SU(2)$. Only those with integer j are representations of $SO(3)$.

Note: All representations of G_0 can be found from those of L. For let G_0' be a representation of G_0, then the Lie algebra of G_0' is homomorphic to L (as we saw earlier).

The representation Γ obtained by exponentiating the representation γ of the Lie algebra is IRR iff γ is IRR.

Note: Let $\bar{G}_0$ be the UCG with Lie algebra $\{\ell\}$. Let $\Gamma = \{f(\ell)\}$ be a representation of $\{\ell\}$ with ℓ represented by $f(\ell)$ in γ. Let Γ be the

113

associated group. The homomorphism $\bar{G}_0 \to \Gamma$ is then given by

$$e^{t\ell} \to e^{tf(\ell)},$$

$$\prod_i e^{t_i \ell_i} \to \prod_i e^{t_i f(\ell_i)}.$$

Here both sides should be ordered in the same way in the products. $e^{t\ell}$ has a clear meaning for matrix groups. For its meaning for abstract groups, we refer the reader to Pontrjagin's book.

ii) The Lie algebra $L_{SO(3)}$ of $SO(3)$

If $\{g(t)\}$ is a one-parameter subgroup, $g(t) = e^{tu}$ for some u, then

1) $g(t)^T g(t) = 1 = e^{tu^T} e^{tu}$. Differentiate in t and set $t = 0$. Then $u + u^T = 0$ implies that u is antisymmetric.

2) Since $g(t)$ is real, $\dot{g}(t)|_{t=0} = u$ is real. Conversely, e^{tu} for u real and antisymmetric is real and orthogonal.

3) Det $e^{tu} = 1$ or -1. However, det $e^{tu} = -1$ cannot occur because e^{tu} is continuous in t and hence constant in t, and it is 1 as $t \to 0$. Thus

$$L_{SO(3)} = \{u|u \text{ real}, \ u + u^T = 0\}.$$

A basis is $L_i, i = 1, 2, 3$, $(L_i)_{jk} = -\varepsilon_{ijk}$:

$$L_1 = -\begin{bmatrix} 0 & 0 & 0 \\ 0 & 0 & 1 \\ 0 & -1 & 0 \end{bmatrix}, \quad L_2 = -\begin{bmatrix} 0 & 0 & -1 \\ 0 & 0 & 0 \\ 1 & 0 & 0 \end{bmatrix}, \quad L_3 = -\begin{bmatrix} 0 & 1 & 0 \\ -1 & 0 & 0 \\ 0 & 0 & 0 \end{bmatrix}.$$

The dimension of $L_{SO(3)} = \dim SO(3) = 3$. The Lie bracket is given by

$$[L_i, L_j] = \varepsilon_{ijk} L_k. \qquad (*)$$

The abstract Lie algebra (which is the relevant object for representation theory, etc.) is defined by a real vector space with basis L_i and Lie bracket $(*)$. The other properties of the matrices L_i above are not relevant.

Define $M_{\mu\nu}$, $\mu, \nu = 1, 2, 3$ as follows:

$$(M_{\mu\nu})_{\rho\sigma} = \delta_{\mu\rho}\,\delta_{\nu\sigma} - \delta_{\mu\sigma}\,\delta_{\nu\rho}.$$

Then

$$L_1 = -M_{23}, \ L_2 = -M_{31}, \ L_3 = -M_{12}$$

and

$$[M_{\mu\nu}, M_{\rho\sigma}] = \delta_{\nu\rho} M_{\mu\sigma} + \delta_{\mu\sigma} M_{\nu\rho} - \delta_{\mu\rho} M_{\nu\sigma} - \delta_{\nu\sigma} M_{\mu\rho}.$$

This information is useful for the generalization from $SO(3)$ to $SO(n)$.

iii) The Lie algebra $L_{SU(n)}$ of $SU(n)$

Consider a one-parameter subgroup $g(t) = e^{tu} \in SU(n)$. Then
1) $g^\dagger g = 1 = e^{tu^\dagger} e^{tu}$ or $u^\dagger + u = 0$.
2) Det $g = \det e^{tu} = 1$ or tr $u = 0$.

Let $A_\sigma^\rho, \rho, \sigma = 1, 2, \ldots, n$ be matrices with
$$(A_\sigma^\rho)_{ij} = \delta_{\rho i} \delta_{\sigma j}.$$
We can write $u = \varepsilon_\sigma^\rho A_\sigma^\rho$. Then 2) implies that tr $u = \Sigma \varepsilon_i^i = 0$. In order to avoid constraints on ε_σ^ρ, define
$$B_\sigma^\rho \equiv A_\sigma^\rho - \frac{1}{n} \delta_\sigma^\rho 1.$$
Then tr $B_\sigma^\rho = $ tr $A_\sigma^\rho - \delta_\sigma^\rho = \delta_\sigma^\rho - \delta_\sigma^\rho = 0$. Thus if we write $u = \varepsilon_\sigma^\rho B_\sigma^\rho$, then tr $u = \varepsilon_\sigma^\rho$ tr $B_\sigma^\rho = 0$ for any ε_σ^ρ . Note that $B_\sigma^\rho (\rho \neq \sigma)$ are linearly independent, but
$$\Sigma B_\rho^\rho = \Sigma A_\rho^\rho - 1 = 0$$
implies that B_ρ^ρ $(\rho = 1, 2, \cdots, n)$ are not linearly independent.

Example:

$SU(3)$:

$$B_1^1 = \begin{bmatrix} 1\,0\,0 \\ 0\,0\,0 \\ 0\,0\,0 \end{bmatrix} - \frac{1}{3}1 = \begin{bmatrix} 2/3 & 0 & 0 \\ 0 & -1/3 & 0 \\ 0 & 0 & -1/3 \end{bmatrix},$$

$$B_2^2 = \begin{bmatrix} -1/3 & 0 & 0 \\ 0 & 2/3 & 0 \\ 0 & 0 & -1/3 \end{bmatrix}, \quad B_3^3 = \begin{bmatrix} -1/3 & 0 & 0 \\ 0 & -1/3 & 0 \\ 0 & 0 & 2/3 \end{bmatrix}.$$

Condition 1) implies $(B_\sigma^{\rho\dagger} = B_\rho^\sigma)$:
$$0 = \varepsilon_\sigma^\rho B_\sigma^\rho + \varepsilon_\sigma^{\rho*} B_\rho^\sigma$$
$$= \varepsilon_\sigma^\rho B_\sigma^\rho + \varepsilon_\rho^{\sigma*} B_\sigma^\rho.$$
Since B_σ^ρ are linearly independent for $\rho \neq \sigma$, one has
$$\varepsilon_\sigma^\rho + \varepsilon_\sigma^{\rho*} = 0.$$
For $\rho = \sigma$, 1) implies
$$\sum_\rho (\varepsilon_\rho^\rho - \varepsilon_\rho^{\rho*}) B_\rho^\rho = 0$$

or

$$\sum_{i=1}^{n-1} \mathrm{Re}(\varepsilon_i^i - \varepsilon_n^n)B_i^i = 0.$$

Using the linear independence of B_i^i, $i = 1, 2, \ldots, n-1$ one finds:

$$\mathrm{Re}\,\varepsilon_1^1 = \mathrm{Re}\,\varepsilon_2^2 = \ldots = \mathrm{Re}\,\varepsilon_n^n.$$

Thus an element of $L_{SU(n)}$ is of the form $\varepsilon_\sigma^\rho B_\sigma^\rho$ where

1) $\varepsilon_\sigma^\rho + \varepsilon_\rho^{\sigma*} = 0$ for $\rho \neq \sigma$,
2) $\mathrm{Re}\,\varepsilon_1^1 = \mathrm{Re}\,\varepsilon_2^2 = \ldots = \mathrm{Re}\,\varepsilon_n^n$,
3) $[B_\sigma^\rho,\, B_\nu^\mu] = \delta_\nu^\rho B_\nu^\mu - \delta_\sigma^\mu B_\nu^\rho$.

We now compute a basis for $L_{SU(n)}$:

$$\varepsilon_\sigma^\rho B_\sigma^\rho = \sum_\sigma \sum_{\rho \leq \sigma-1} \varepsilon_\sigma^\rho B_\sigma^\rho + \sum_\sigma \sum_{\rho \geq \sigma+1} \varepsilon_\sigma^\rho B_\sigma^\rho + \sum_\rho \varepsilon_\rho^\rho B_\rho^\rho.$$

In the second term, rewrite the summation as $\sum_\rho \sum_{\sigma \leq \rho-1}$ and interchange indices, then

$$\sum_{\rho,\sigma} \varepsilon_\sigma^\rho B_\sigma^\rho = \sum_\sigma \sum_{\rho \leq \sigma-1} [\varepsilon_\sigma^\rho B_\sigma^\rho + \varepsilon_\rho^\sigma B_\rho^\sigma] + i \sum_{\rho \leq n-1} \mathrm{Im}(\varepsilon_\rho^\rho - \varepsilon_n^n)B_\rho^\rho,$$

$$= \sum_\sigma \sum_{\rho \leq \sigma-1} [\varepsilon_\sigma^\rho B_\sigma^\rho - \varepsilon_\sigma^{\rho*} B_\rho^\sigma] + i \sum_{\rho \leq n-1} \mathrm{Im}(\varepsilon_\rho^\rho - \varepsilon_n^n)B_\rho^\rho.$$

Now write

$$\varepsilon_\sigma^\rho B_\sigma^\rho - \varepsilon_\sigma^{\rho*} B_\rho^\sigma = (\varepsilon_\sigma^\rho - \varepsilon_\sigma^{\rho*})\left(\frac{B_\sigma^\rho + B_\rho^\sigma}{2}\right) + (\varepsilon_\sigma^\rho + \varepsilon_\sigma^{\rho*})\left(\frac{B_\sigma^\rho - B_\rho^\sigma}{2}\right),$$

$$= 2i\,\mathrm{Im}\varepsilon_\sigma^\rho\left(\frac{B_\sigma^\rho + B_\rho^\sigma}{2}\right) + 2\,\mathrm{Re}\,\varepsilon_\sigma^\rho\left(\frac{B_\sigma^\rho - B_\rho^\sigma}{2}\right).$$

Thus

$$\sum_{\rho,\sigma} \varepsilon_\sigma^\rho B_\sigma^\rho = \sum_\sigma \sum_{\rho \leq \sigma-1} \left[2i\,\mathrm{Im}\varepsilon_\sigma^\rho\left(\frac{B_\sigma^\rho + B_\rho^\sigma}{2}\right) + 2\,\mathrm{Re}\,\varepsilon_\sigma^\rho\left(\frac{B_\sigma^\rho - B_\rho^\sigma}{2}\right)\right]$$
$$+ \sum_{\rho \leq n-1} \mathrm{Im}(\varepsilon_\rho^\rho - \varepsilon_n^n)iB_\rho^\rho.$$

By writing in this way we eliminate the constraints on ε_σ^ρ. A basis for the real Lie algebra of $SU(n)$ is thus:

$$-i\left(\frac{B_\sigma^\rho + B_\rho^\sigma}{2}\right),\quad -\left(\frac{B_\sigma^\rho - B_\rho^\sigma}{2}\right),$$

with

$$\rho = 1, 2, \ldots, n-1;\ \sigma = 2, 3, \ldots, n; \rho \leq \sigma - 1;$$

and

$$-iB_1^1, \; -iB_2^2, \ldots, -iB_{n-1}^{n-1}.$$

The dimension of $L_{SU(n)}$ is :

$$2(1 + 2 + \ldots + n - 1) + n - 1 = n(n-1) + n - 1 = n^2 - 1 = \dim SU(n).$$

Example:

$SU(2)$:

$$-iB_1^1 = -i\left\{\begin{bmatrix} 1 & 0 \\ 0 & 0 \end{bmatrix} - \frac{1}{2}\begin{bmatrix} 1 & 0 \\ 0 & 1 \end{bmatrix}\right\} = -i\begin{bmatrix} 1/2 & 0 \\ 0 & -1/2 \end{bmatrix} = -i\frac{\sigma_3}{2} \equiv \ell_3,$$

$$-i\left(\frac{B_2^1 + B_1^2}{2}\right) = -\frac{i}{2}\begin{bmatrix} 0 & 1 \\ 1 & 0 \end{bmatrix} = -i\frac{\sigma_1}{2} \equiv \ell_1,$$

$$-\left(\frac{B_2^1 - B_1^2}{2}\right) = -\frac{1}{2}\begin{bmatrix} 0 & 1 \\ -1 & 0 \end{bmatrix} = -i\frac{\sigma_2}{2} \equiv \ell_2.$$

Thus

$$[\ell_i, \ell_j] = \varepsilon_{ijk}\ell_k$$

and one sees that $L_{SO(3)}$ and $L_{SU(2)}$ are isomorphic.

Lemma

Let $L = \{\ell\}$ be a Lie algebra and $\gamma = \{\gamma(\ell)\}$ be a representation with associated Lie group Γ of linear operators. Then Γ is unitary iff γ is anti-hermitian.

The proof follows by observing that any element in Γ can be written in the form

$$e^{t_1\gamma(\ell_1)} \, e^{t_2\gamma(\ell_2)} \, \cdots$$

It is customary in physics to define $\tilde{\gamma} = i\gamma$ and look at hermitian representations of $\tilde{\gamma}$.

Example:

$SO(3)$:
We define $J_i = iL_i$. In a unitary representation, $J_i^\dagger = J_i$ with

$$[J_i, \; J_j] = i\varepsilon_{ijk}J_k.$$

In the (3×3)-dimensional representation, $(J_i)_{jk} = -i\varepsilon_{ijk}$.

LECTURE 23

i) The center of a Lie algebra

Consider a Lie algebra L. The center $\mathcal{C}$ of L is defined as the set

$$\{c \in L | [c, x] = 0, \ \forall \ x \in L\}.$$

In any IRR of L, elements c of $\mathcal{C}$ are multiples of the identity, $\lambda_c 1$. Their values can thus be used to label the IRR's.

Example:

Heisenberg algebra: This algebra has a basis P, Q, and e where

$$[Q, \ P] = e, \ [P, \ P] = [Q, \ Q] = [P, \ e] = [Q, \ e] = 0.$$

The center is

$$\mathcal{C} = \{\lambda e | \lambda \text{ real}\}.$$

The algebra can be constructed from the familiar operators in quantum theory:

$$Q = iq, \ P = ip, \ e = i\hbar.$$

One has the bracket relation $[q, \ p] = i\hbar$ with all other brackets vanishing. $\hbar$ is chosen to be Planck's constant$/2\pi$ in quantum theory.

ii) Casimir operators

Definition. Let L be a Lie algebra with basis L_i, $i=1, 2, \ldots, n$. A Casimir operator is a formal polynomial in the L_i which commutes with all elements of L. [More precisely, it can be defined as a central element in the "enveloping algebra" of L.]

Examples:

1) $SU(2)$ or $SO(3)$. With J_i as the basis, one forms

$$C_2 = \sum_i J_i \, J_i \equiv J^2.$$

Then

$$[C_2, \, J_i] = 0.$$

In an IRR, C_2 is a multiple of the identity, namely $j(j+1)\mathbf{1}$. Consequently we can label the IRR's by the eigenvalues of C_2.

2) $SU(3)$. There are two independent Casimir operators for $SU(3)$:

$$C_2 = B^\rho_\lambda B^\lambda_\rho$$
$$C_3 = B^\rho_\lambda B^\lambda_\sigma B^\sigma_\rho.$$

We have

$$[C_i, \, B^\lambda_\rho] = 0, \quad i = 2, 3.$$

iii) General considerations on the adjoint representation

Let L be a real Lie algebra with basis L_i, $i=1, 2, \ldots, n$ and Lie bracket $[L_i, L_j] = C^k_{ij} L_k$. The structure constants C^k_{ij} are real, $C^k_{ij} = -C^k_{ji}$, and they define a representation of the Lie algebra via $L_i \to D(L_i)$ where $D(L_i)_{kj} = C^k_{ij}$. We now discuss this representation in more detail.

The adjoint representation of the Lie algebra

$$\gamma^A = \{\mathrm{ad}\ell | \ell \in L\}$$

is defined by

$$\mathrm{ad}\ell \, x = [\ell, \, x], \quad x \in L.$$

The matrix of $\mathrm{ad}L_i$ in the basis $L_1, L_2, \ldots, L_n$ will be called $D(L_i)$. Then

$$\mathrm{ad}L_i L_j = D(L_i)_{kj} L_k = [L_i, \, L_j] = C^k_{ij} L_k$$

or

$$D(L_i)_{kj} = C_{ij}^k.$$

The associated representation of the group G_0 is the adjoint representation $\Gamma^A = \{\text{Ad } g\}$ defined by

$$g = e^{\theta_1 L_1} e^{\theta_2 L_2} \ldots \to \text{Ad } g = e^{\theta_1 \text{ad } L_1} e^{\theta_2 \text{ad } L_2} \ldots .$$

The matrix of Ad g is

$$\mathcal{D}(g) = e^{\theta_1 D(L_1)} e^{\theta_2 D(L_2)} \ldots .$$

Below we list some properties of Ad g.

1) Ad g is the linear operator on L defined by

$$\text{Ad } g \, x = g \, x \, g^{-1}$$

where, for example,

$$e^{\theta_i L_i} x e^{-\theta_i L_i} = \sum_{k=0}^{\infty} \frac{\theta_i \theta_j \ldots \theta_k}{k!} [L_i, [L_j, \ldots [L_k, x] \ldots]].$$

For the right-hand side is easily seen to be $e^{\theta_i \text{ ad } L_i} x$.

2) The equation $g \, L_i \, g^{-1} = \mathcal{D}(g)_{ij} L_j$ is true in any representation of G_0 and L. [That is, g and L_i can stand for operators in any representation.] [Prove this as an exercise.]

3) $D(g \, x \, g^{-1}) = \mathcal{D}(g) D(x) \mathcal{D}(g^{-1})$

This follows from 2) and 3).

Examples:

1) $SO(3)$. The Lie algebra (in the defining representation) has the basis $\{L_i\}$ where $(L_i)_{jk} = -\varepsilon_{ijk}$. If $R \in$ the defining representation of $SO(3)$, then we know from quantum mechanics that

$$\text{Ad } R \, L_i = R L_i R^{-1} = R_{ji} L_j.$$

One sees that $\Gamma^A(SO(3))$ is the $j = 1$ representation.

2) $SU(2)$. The Lie algebra is spanned by the matrices $L_i = -i\frac{\sigma_i}{2}$. If $g \in SU(2)$, then

$$\text{Ad} g \, L_i = g L_i g^{-1} = R(g)_{ji} L_j$$

where $R(g)$ is an element of $SO(3)$. $\Gamma^A(SU(2))$ is also the $j = 1$ representation.

iv) The Cartan-Killing form

For x, $y \in L$, define a "scalar product" on L by

$$\langle x, y \rangle \equiv \mathrm{tr}[(\mathrm{ad}x)(\mathrm{ad}y)].$$

$\langle \cdot, \cdot \rangle$ has the following properties:
   1) $\langle x, y \rangle = \langle \mathrm{Ad}g\ x, \mathrm{Ad}g\ y \rangle$,
   2) $\langle \mathrm{ad}L_i\ x, y \rangle + \langle x, \mathrm{ad}L_i\ y \rangle = 0$.
The proof of 1) is easily seen from the relation 3) above and 2) follows as the infinitesimal form of 1).

Define a symmetric matrix g, the Cartan-Killing matrix, by

$$g_{\alpha\beta} = \mathrm{tr}(\mathrm{ad}L_\alpha)(\mathrm{ad}L_\beta).$$

Clearly,

$$g_{\alpha\beta} = (\mathrm{Ad}s)_{\alpha'\alpha}(\mathrm{Ad}s)_{\beta'\beta}g_{\alpha'\beta'}, \ \forall\ s \in G_0,$$

where now $\mathrm{Ad}\ s = \mathcal{D}(s)$. Suppose that g is nondegenerate. Then g^{-1} exists, and writing $g = (\mathrm{Ad}\ s)^T g(\mathrm{Ad}\ s)$, from above, we find

$$g^{-1} = (\mathrm{Ad}\ s^{-1}g^{-1}(\mathrm{Ad}\ s^{-1})^T)$$

for any $s \in G_0$. If we denote $(g^{-1})_{\alpha\beta}$ by $g^{\alpha\beta}$, then we can construct invariant polynomials on L, as for example, $C_2 = g^{\alpha\beta}L_\alpha L_\beta$. To check invariance, note that

$$sC_2s^{-1} = g^{\alpha\beta}(\mathrm{Ad}\ s)_{\alpha'\alpha}(\mathrm{Ad}\ s)_{\beta'\beta}L_{\alpha'}L_{\beta'}$$
$$= [(\mathrm{Ad}\ s)g^{-1}(\mathrm{Ad}\ s)^T]_{\alpha'\beta'}L_{\alpha'}L_{\beta'}.$$

Since this relation is true for any s, set $s = s^{-1}$ and use the above equation to find

$$s^{-1}C_2s = (g^{-1})_{\alpha'\beta'}L_{\alpha'}L_{\beta'} = C_2.$$

Since C_2 commutes with every element of G_0, it commutes with every element of L. Consequently C_2 is a Casmir operator, the quadratic Casimir operator.

Example:

$SO(3)$: The Cartan-Killing matrix is given by

$$g_{ij} = \mathrm{Tr}(\mathrm{ad}L_i)(\mathrm{ad}L_j) = \sum_{m.n}(\mathrm{ad}L_i)_{mn}(\mathrm{ad}L_j)_{nm} = \sum_{m,n}\varepsilon_{imn}\varepsilon_{jnm} = -2\delta_{ij}.$$

So

$$(g^{-1})_{ij} = -\frac{1}{2}\delta_{ij}$$

and the Casmir operator for $SO(3)$ is

$$C_2 = -\frac{1}{2}\sum_i L_i^2.$$

In general, an mth order Casimir operator is given, as is easily shown, by

$$C_m = \text{Tr}[\text{ad}L_{i'_1}\,\text{ad}L_{i'_2}\,\ldots\,\text{ad}L_{i'_m}]g^{i'_1 i_1}\,\ldots\,g^{i'_m i_m}L_{i_1}\,\ldots\,L_{i_m}.$$

The matrices $\text{ad}L_i$ in the trace above can be replaced by any of their representatives in any representation [cf. B. Gruber and L. O'Raifeartaigh, Journal of Mathematical Physics 5, 1796 (1964)]. This procedure may give new results for invariants.

Definition. A subalgebra L_0 of a Lie algebra L is an invariant subalgebra if $[L,\,L_0] \subseteq L_0$.

Example:

$G_0 = SU(2) \times SU(2)$. The Lie algebra L_{G_0} of G_0 is a direct sum of $SU(2)$ Lie algebras,

$$L_{G_0} = L_{SU(2)}^{(1)} \oplus L_{SU(2)}^{(2)}$$

with

$$[L_a^{(1)},\,L_b^{(2)}] = 0, \qquad L_\alpha^{(i)} \in L_{SU(2)}^{(i)}.$$

Either of these $L_{SU(2)}^{(i)}$ is an invariant subalgebra isomorphic to $L_{SU(2)}$.

v) Simple and semi-simple Lie algebras

A Lie algebra is *simple* if it has no invariant subalgebras. It is *semi-simple* if it has no Abelian invariant subalgebras.

Example:

$L_{SU(2)}$ is simple, and $L_{SU(2)\times SU(2)}$ is semi-simple.

Definition. If G^i are connected Lie groups and there is a homomorphism $\tau : G^{(1)} \to G^{(2)}$ with a discrete kernel, $G^{(1)}$ and $G^{(2)}$ are said to be locally isomorphic.

Examples:

1) $G^{(1)} = SU(2)$, $G^{(2)} = SO(3)$.
2) $G^{(1)} = SU(2) \times SU(2)$, $G^{(2)} = SO(4)$.

If a Lie group $G^{(1)}$ is locally isomorphic to a simple (semi-simple) Lie group $G^{(2)}$, then the Lie algebra of $G^{(1)}$ is simple (semi-simple).

Cartan has proved that the matrix g with elements $g_{\alpha\beta}$ is nonsingular if and only if the Lie algebra is semi-simple. It can be shown that a semi-simple Lie algebra is the direct sum of simple Lie algebras.

Example:

The semi-simple Lie algebra $L_{SU(2) \times SU(2)}$ [or $L_{SO(4)}$] is the direct sum $L_{SU(2)} \oplus L_{SU(2)}$ where $L_{SU(2)}$ is simple.

LECTURE 24

i) Compact and simple Lie algebras

Theorem

Let G_0 be a connected compact Lie group with Lie algebra L. Then

a) There exists a basis $\{\hat{L}_i\}$ for L such that if

$$[\hat{L}_i, \hat{L}_j] = \hat{\mathcal{C}}^k_{ij} \hat{L}_k,$$

the structure constants $\hat{\mathcal{C}}^k_{ij}$ in this basis being totally antisymmetric in all three indices.

b) If L is simple as well, and if γ is any representation of L, then in this basis,

$$\mathrm{Tr}\, \gamma(\hat{L}_i)\gamma(\hat{L}_j) = k_\gamma \delta_{ij}.$$

Here $\gamma(\hat{L}_i)$ is the representative of $\hat{L}_i$ in γ and k_γ is a constant which can depend on the representation γ.

For proof, we can proceed as follows: Let $\{L_i\}$ be any basis for the (real) Lie algebra L. The structure constants $\mathcal{C}^k_{ij}$ defined by

$$[L_i, L_j] = \mathcal{C}^k_{ij} L_k$$

are real, therefore $\mathrm{ad}\, L$ maps L onto itself. The associated adjoint group $\mathrm{Ad}\, G_0$ thus also does the same. This means that if $s \in G_0$ and

$$s L_i s^{-1} = \mathcal{D}(s)_{ji} L_j,$$

then $\mathcal{D}(s)$ is *real*.

Since G_0 is compact, there is the basis $\{\hat{L}_i\}$,

$$\hat{L}_i = [\int_{G_0} d\mu(s)\ \mathcal{D}(s)^\dagger \mathcal{D}(s)]^{-1/2}_{ji} L_j$$

125

[see Lectures 9, 15] relative to which the matrix of s is unitary:

$$s\hat{L}_i s^{-1} = \hat{\mathcal{D}}(s)_{ji}\hat{L}_j, \quad \hat{\mathcal{D}}(s)^\dagger = \hat{\mathcal{D}}(s)^{-1} = \hat{\mathcal{D}}(s^{-1}).$$

Since $\mathcal{D}(s)$ is real, the formula above shows that the transformation connecting $\{L_i\}$ to $\{\hat{L}_i\}$ is real. Therefore it follows that $\{\hat{L}_i\}$ is a basis for the real Lie algebra L and $\hat{\mathcal{D}}(s)$ is real. It is thus *real* and *orthogonal* :

$$\hat{\mathcal{D}}(s)^* = \hat{\mathcal{D}}(s), \quad \hat{\mathcal{D}}(s)^T \hat{\mathcal{D}}(s) = 1.$$

Now if we write

$$\text{ad } \hat{L}_i \, \hat{L}_j = \hat{\mathcal{C}}_{ij}^k \hat{L}_k \equiv \hat{D}(\hat{L}_i)_{kj}\hat{L}_k,$$

then $\hat{\mathcal{D}}$ is of the form

$$\exp[t_1 \hat{D}(\hat{L}_1)] \exp[t_2 \hat{D}(\hat{L}_2)] \ldots$$

where t_i are real. Thus $\hat{D}(\hat{L}_i)$ are real and antisymmetric:

$$\hat{D}(\hat{L}_i)^* = \hat{D}(\hat{L}_i), \quad \hat{D}(\hat{L}_i)^T = -\hat{D}(\hat{L}_i).$$

Since

$$\hat{D}(\hat{L}_i)_{kj} = \hat{\mathcal{C}}_{ij}^k,$$

$\hat{\mathcal{C}}_{ij}^k$ is (real and) antisymmetric in k and j. Since we already know that it is antisymmetric in i and j, it is totally antisymmetric. This proves a).

To prove b), we need the result that the representation ad L for a simple L is irreducible. This is proved as follows: Let $L_0 \subseteq L$ be an invariant subspace under ad L. Then symbolically

$$\text{ad } L \, L_0 = [L, \, L_0] \subseteq L_0.$$

Therefore, L_0 is an invariant subalgebra. Since L is simple, $L_0 = \{0\}$ or L and the result is proved. Thus if L is simple, $\text{Ad} G_0$ is irreducible.

Let Γ be the group associated with γ and let $\Gamma(s)$ be the representative of $s \in G_0$ in Γ. We know that

$$\Gamma(s)\gamma(\hat{L}_i)\Gamma(s)^{-1} = \hat{\mathcal{D}}(s)_{ji}\gamma(\hat{L}_j)$$

and hence we infer that

$$\text{Tr } \gamma(\hat{L}_i)\gamma(\hat{L}_j) \equiv \hat{g}_{ij}$$
$$= \text{Tr } \{\Gamma(s)\gamma(\hat{L}_i)\Gamma(s)^{-1}\}\{\Gamma(s)\gamma(\hat{L}_j)\Gamma(s)^{-1}\}$$
$$= \hat{\mathcal{D}}(s)_{ki} \, \hat{\mathcal{D}}_{\ell j}\hat{g}_{k\ell}.$$

In view of the orthogonality of $\hat{\mathcal{D}}(s)$,

$$[\hat{\mathcal{D}}(s), \, \hat{g}] = 0.$$

In view of Schur's lemma, the result b) follows.

Remark.

We can evaluate k_γ by setting $i = j$ and summing:

$$\text{Tr}(\sum_i \gamma(\hat{L}_i)\gamma(\hat{L}_i)) = k_\gamma \mid \text{ad } L \mid,$$

$$\mid \text{ad } L \mid = \text{ dimension of } L.$$

Suppose γ is IRR. Then $\Sigma\gamma(\hat{L}_i)\gamma(\hat{L}_i)$, the quadratic Casimir operator, is a multiple of identity:

$$\sum_i \gamma(\hat{L}_i)\gamma(\hat{L}_i) = C_\gamma^{(2)}1.$$

Thus

$$k_\gamma = \frac{\mid \gamma \mid}{\mid \text{ad } L \mid}C_\gamma^{(2)},$$

$$\mid \gamma \mid = \text{ dimension of } \gamma.$$

Example:

Let $L = L_{SU(2)}$ with basis $\left\{\gamma(\hat{L}_i) = -i\frac{\sigma_i}{2}\right\}$ in the two-dimensional representation γ. It fulfills

$$\text{Tr } \gamma(\hat{L}_i)\gamma(\hat{L}_j) = -\frac{1}{2}\delta_{ij}$$

as an explicit computation shows. We show now that the formula above also gives $-\frac{1}{2}$ for k_γ. We have

$$\sum_i \gamma(\hat{L}_i)\gamma(\hat{L}_i) = -\frac{3}{4}1,$$

or

$$C_\gamma^{(2)} = -\frac{3}{4}.$$

Since $\mid \gamma \mid = 2, \mid \text{ad}L \mid = 3$, we find $k_\gamma = -\frac{1}{2}$.

LECTURE 25

i) The rank of a Lie algebra

Definition. The rank of a Lie algebra L is the maximum number of linearly independent elements in L which commute with each other. [In Dirac's terminology, these elements form a complete commuting set (CCS) of operators in L.]

Examples:

1) $SU(2)$ or $SU(3)$: Any one of the L_i gives a CCS, so the rank is one.

2) $SU(3)$: If we take B_σ^ρ as generators of $L_{SU(3)}$, then a CCS is $\{B_1^1,\ B_2^2\}$. The rank is two.

3) $SU(n)$: A generalization of a CCS for $SU(3)$ gives the rank as $n-1$.

The vector space spanned by a CCS is a Cartan subalgebra. It is unique upto conjugation.

In a given IRR, we can simultaneously diagonalize a CCS and use their eigenvalues to label the states (at least partially).

Example:

$SU(2)$: A CCS is $\{L_3\}$ and $L_3|jm\rangle = -im|jm\rangle$ for the usual basis $|jm\rangle$.

It can be shown that the number of independent Casimir operators for a semi-simple Lie algebra is the rank of the Lie algebra.

Example:

$SU(n)$: Rank $= n - 1$. The independent Casimir operators are:

$$C_2 = B^\rho_\sigma B^\sigma_\rho, \quad C_3 = B^\rho_\sigma B^\sigma_\lambda B^\lambda_\rho, \ldots,$$

$$C_n = \underbrace{B^\rho_\sigma B^\sigma_\lambda \ldots B^\epsilon_\rho}_{n \text{ factors}}.$$

ii) The irreducible representations of $SO(3)$ or $SU(2)$ Lie algebras

In the following we use the hermitian form of the generators: $J_i = iL_i$ and $[J_i, J_j] = i\varepsilon_{ijk}J_k$. We first list some general properties of representations for these groups.

1) Since $SU(2)$ or $SO(3)$ is compact, the IRR's are finite dimensional.

2) We can assume that the representation is unitary, i.e., $J_i^\dagger = J_i$.

3) A CCS is J_3. Since $J_3^\dagger = J_3$, it can be diagonalized, and its eigenvectors give a basis for the carrier space of the IRR.

Let V be the complex vector space on which the IRR is realized. Let $|m\rangle$ be a basis for V with $J_3|m\rangle = m|m\rangle$, m real and $\langle m'|m\rangle = \delta_{m',m}$. Since dim V is finite, $|m|$ is bounded. We define max $m = j$ and min $m = j - n_0$.

Defining $J_\pm = J_1 \pm iJ_2$, we have the commutation relations

$$[J_3, J_\pm] = \pm J_\pm, \quad [J_+, J_-] = 2J_3.$$

Using these relations, one has

$$J_3 J_\pm|m\rangle = (m \pm 1)J_\pm|m\rangle,$$

and we can conclude that

$$J_+|m\rangle = \rho_m|m+1\rangle, \quad J_-|m\rangle = \sigma_m|m-1\rangle.$$

Because j and $j - n_0$ are the maximum and minimum eigenvalues of J_3 in this representation,

$$J_+|J\rangle = 0 \quad \text{or} \quad \rho_j = 0,$$

$$J_-|j - n_0\rangle = 0 \quad \text{or} \quad \sigma_{j-n_0} = 0.$$

We can show that n_0 must be an integer as follows: Because V is finite dimensional, there must exist an integer $\bar{n}_0$ such that $(J_-)^{\bar{n}_0}|j\rangle \neq 0$ and $(J_-)^{\bar{n}_0+1}|j\rangle = 0$. But this implies that $\sigma_{j-\bar{n}_0} = 0$, so $\bar{n}_0 = n_0 =$ integer.

The numbers ρ_m and σ_m can be derived from the commutation relations. Use the expectation value

$$\langle m|[J_+, J_-]|m\rangle = 2m$$

and the definition of ρ_m, σ_m to conclude that

$$\rho_{m-1}\sigma_m - \rho_m\sigma_{m+1} = 2m. \quad (A)$$

From the hermiticity of the J_i, $(J_+)^\dagger = J_-$, so

$$\langle m|J_+|n\rangle^* = \langle n|J_-|m\rangle,$$

or

$$\rho^*_{m-1} = \sigma_m. \quad (B)$$

We can solve (A, B) with $\rho_j = \sigma_{j-n_0} = 0$, n_0 a non-negative integer by substituting (B) into (A). Then

$$|\rho_{m-1}|^2 - |\rho_m|^2 = 2m.$$

Staring with $m = j$, we get

$$|\rho_{j-1}|^2 = 2j, \quad |\rho_{j-2}|^2 = 2(j-1) + 2j, \text{ etc.}$$

Consequently,

$$|\rho_{j-m}|^2 = 2[j - (m-1)] + \ldots + 2j = 2mj - \frac{2m(m-1)}{m} = m(2j - m + 1)$$

or

$$|\rho_m|^2 = (j - m)(j + m + 1).$$

From (B), we have

$$|\sigma_m|^2 = (j + m)(j - m + 1).$$

If we check the boundary conditions, we see that $\rho_j = 0$ is satisfied while

$$|\sigma_{j-n_0}|^2 = 0 = (2j - n_0)(n_0 + 1)$$

implies that

$$n_0 = 2j$$

or that

$$j = \frac{n_0}{2}, \quad n_0 = 0, 1, 2, \ldots.$$

Thus V is spanned by $\{|m\rangle\}$, $m = -j, -j+1, \ldots, +j$ where in a given IRR, j is integral or half integral. Summarizing:

$$\langle m'|m\rangle = \delta_{m',m},$$
$$J_3|m\rangle = m|m\rangle,$$
$$J_+|m\rangle = \rho_m|m+1\rangle, \quad |\rho_m|^2 = (j-m)(j+m+1),$$
$$J_-|m\rangle = \sigma_m|m-1\rangle, \quad |\sigma_m|^2 = (j+m)(j-m+1),$$
$$\rho_m^* = \sigma_{m-1}.$$

Since the overall phase of a state is undetermined, we may use this freedom to remove the ambiguity in the phases of ρ_m, σ_m. Thus let ρ_m be the positive square root of $|\rho_m|^2$. Then by changing the phases of $|m\rangle$, we can set $J_+|m\rangle = \rho_m|m+1\rangle$. Then $\rho^*_m = \sigma_{m-1}$ implies that σ_m is also real and non-negative. With this phase convention, the matrices $J_3, J_\pm$ are real, i.e. J_1 and J_3 are real, and J_2 is pure imaginary.

Irreducibility of These $SU(2)$ or $SO(3)$ Lie Algebra Representations

Denote the spin j representation of $L_{SU(2)} = L_{SO(3)}$ by $\gamma^{(j)}$. Let $\bar{V} \subseteq V^{(j)}$ be a nontrivial invariant subspace under $\gamma^{(j)}$. Since $\bar{V}$ is invariant under the hermitian J_3, we can diagonalize J_3 in $\bar{V}$. Let j_0 be the maximum eigenvalue of J_3 in $\bar{V}$ with $J_3|j_0\rangle = j_0|j_0\rangle$. That j_0 is the maximum eigenvalue implies that $J_+|j_0\rangle = 0$. However, $J_+|j_0\rangle = \rho_{j_0}|j_0+1\rangle$ and ρ_{j_0} does not vanish unless $j_0 = j$, so $|j\rangle \in \bar{V}$. Since $|j\rangle \in \bar{V}, (J_-)^m|j\rangle \in \bar{V}$ as well, and $\sigma_m \neq 0$ unless $m = -j$. Thus all the states $|j\rangle, |j-1\rangle, \ldots, |-j\rangle \in \bar{V}$. This implies that $\bar{V} = V^{(j)}$ since these are the basis vectors for $V^{(j)}$.

Remarks.

The representation $\Gamma^{(j)}$ of $SU(2)$ is obtained via the association

$$SU(2) \ni g = e^{\frac{i\sigma_3\gamma}{2}}e^{\frac{i\sigma_2\beta}{2}}e^{\frac{i\sigma_3\alpha}{2}} \to D^{(j)}(g) = e^{iJ_3\gamma}e^{iJ_2\beta}e^{iJ_3\alpha} \in \Gamma^{(j)}.$$

For j not equal to an integer, $e^{i2\pi j_3}|m\rangle = -|m\rangle$, while in $SO(3)$, the corresponding operator is 1. Thus $\Gamma^{(j)}$ is always a representation of $SU(2)$, and a representation of $SO(3)$ if j is an integer.

For matrix elements:

$$D_{mn}^{(j)}(g) = \langle m|D^{(j)}(g)|n\rangle$$
$$= e^{im\gamma}e^{in\alpha}\langle m|e^{i\beta J_2}|n\rangle$$
$$= e^{im\gamma}e^{in\alpha}d_{mn}^{(j)}(\beta).$$

$d^{(j)}(\beta)$ is called the reduced rotation matrix. Since J_2 is pure imaginary in the basis we have chosen, $d^{(j)}_{mn}(\beta)$ is real.

A useful property of the J_i matrices in the IRR $\gamma^{(j)}$ is

$$(J_i)^{2j+1} = \sum_{\nu=0}^{2j} \xi_\nu (J_i)^\nu$$

where ξ_ν are real numbers.

Proof:

J_i has eigenvalues $-j, -j+1, \ldots, +j$. Therefore since a matrix satisfies its own characteristic equation,

$$\prod_{\nu=-j}^{+j} (J_i - \nu) = 0.$$

Expanding the product in powers of J_i, one gets the above result.

LECTURE 26

i) Tensor methods for groups

Let $\Gamma = \{t(g)\}$ be a representation of a group G on a vector space $\mathbb{C}^n = \{x | x = (x^1, \ldots, x^n)\}$ where

$$[t(g)x]^i = D(g)_{ij}x^j.$$

A tensor of rank ν associated with the representation Γ of G is a vector F in an n^ν-dimensional vector space $\mathbb{C}^{n^\nu}$. A component of this vector is denoted by

$$F^{j_1 j_2 \cdots j_\nu}, \quad j_i = 1, 2, \ldots, n.$$

By definition $\mathbb{C}^{n^\nu}$ carries the representation

$$\Gamma^\nu = \underbrace{\Gamma \otimes \Gamma \otimes \ldots \otimes \Gamma}_{\nu \text{ factors}}$$

via $\{T(g)\}$ where

$$[T(g)F]^{j_1 \cdots j_\nu} = D(g)_{j_1 j_1'} D(g)_{j_2 j_2'} \cdots D(g)_{j_\nu j_{\nu'}} F^{j_1' j_2' \cdots j_{\nu'}}.$$

The permutation group S_ν has a natural action on $\mathbb{C}^{n^\nu}$ as follows: If

$$s = \begin{pmatrix} 1 & 2 & \ldots & \nu \\ 1' & 2' & \ldots & \nu' \end{pmatrix},$$

associate to s a linear operator on $\mathbb{C}^{n^\nu}$, also called s, via

$$[sF]^{j_1 \cdots j_\nu} = F^{j_{s(1)} \cdots j_{s(\nu)}}, \quad s(j) \equiv j'.$$

Example:

If $s = (123)$, then

$$(sF)^{\substack{j_1\ j_2\ j_3 \\ 4\ 6\ 5}} = F^{\substack{j_{s(1)}\ j_{s(2)}\ j_{s(3)} \\ 6\ 5\ 4}}$$

since $[s(1),\ s(2),\ s(3)] = [2, 3, 1]$ corresponds to

$$[j_{s(1)},\ j_{s(2)},\ j_{s(3)}] = [6, 5, 4].$$

Definition. A tensor of rank ν is symmetric if $sF = F, \forall s \in S_\nu$. The components of a symmetric tensor which differ only in the order of the indices are equal.

Definition. Let $\hat{\alpha}$ be a linear operator on $\mathbb{C}^{n^\nu}$ defined by

$$[\hat{\alpha}\ F]^{j_1 j_2 \cdots j_\nu} = \alpha_{j_1 j_2 \cdots j_\nu} k_1 k_2 \ldots k_\nu F^{k_1 k_2 \ldots k_\nu}$$

or

$$[\hat{\alpha}\ F]^{(j)} = \alpha_{(j),(k)}\ F^{(k)}.$$

Then we call $\hat{\alpha}$ bisymmetric if $\hat{\alpha}$ commutes with all s. [Here s is the linear operator representing a permutation in S_ν.] Now

$$[\hat{\alpha} s F]^{(j)} = \alpha_{(j),(k)}\ [sF]^{(k)} = \alpha_{(j),(k)}\ F^{s(k)} = \alpha_{(j),s^{-1}(k)}\ F^{(k)},$$
$$[s\hat{\alpha}\ F]^{(j)} = [\hat{\alpha}\ F]^{s(j)} = \alpha_{s(j),(k)}\ F^{(k)}.$$

$[s(k)$ now denotes $(k_{s(1)},\ k_{s(2)},\ \ldots k_{s(\nu)})$.] Thus if $\hat{\alpha}$ is bisymmetric,

$$\alpha_{(j),s^{-1}(k)} = \alpha_{s(j),(k)}$$

or

$$\alpha_{(j),(k)} = \alpha_{s(j),s(k)}.$$

Theorem

The representation Γ^ν, i.e., every linear operator in Γ^ν, is bisymmetric. A typical matrix element of an operator in Γ^ν is

$$D^\nu_{(j),(k)}(g) \equiv D_{j_1 k_1}(g) \cdots D_{j_\nu k_\nu}(g)$$

and the action of s on the lower indices interchanges the factors $D_{j_\rho k_\rho}(g)$. Hence $D_{(j),(k)}(g) = D^\nu_{s(j),s(k)}(g)$. Consequently $s\mathbb{C}^{n^\nu}$ is invariant under Γ^ν:

$$T(g)s\mathbb{C}^{n^\nu} = sT(g)\mathbb{C}^{n^\nu} = s\mathbb{C}^{n^\nu}.$$

If $\mathbb{C}S_\nu$ is the group algebra of S_ν, and $x \in \mathbb{C}S_\nu$, then we have that $x\mathbb{C}^{n^\nu}$ is invariant under Γ^ν for similar reasons. [Hereafter we identify the linear operators $\{s\}$ with S_ν. Thus $\mathbb{C}S_\nu$ is algebra generated by these operators.]

In the analysis of the representations of the following groups, we shall choose the x's in such a way that we are able to reduce Γ^ν as much as possible.

ii) The group $GL(n, \mathbb{C})$

This group is not compact. As a consequence, all of its finite dimensional representations, except the trivial one, are nonunitary. It has infinite dimensional unitary representations. The considerations below give incomplete results for this group.

Reduction of Tensors of Rank ν of $GL(n, \mathbb{C})$

Here Γ is taken to be the defining $n \times n$ representation.

1) Write down the frames of S_ν. For $\nu = 3$, one has

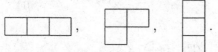

2) Write down the standard tableaux for these frames. For $\nu = 3$, one has

$$\boxed{1\ 2\ 3}\ , \quad \begin{array}{cc}\boxed{1}&\boxed{2}\\ \boxed{3}\end{array}\ , \quad \begin{array}{cc}\boxed{1}&\boxed{3}\\ \boxed{2}\end{array}\ , \quad \begin{array}{c}\boxed{1}\\ \boxed{2}\\ \boxed{3}\end{array}\ .$$

3) For each tableau, let $e = QP$. Then $e\mathbb{C}^{n^\nu}$ carries an IRR of $GL(n, \mathbb{C})$. Further

$$\mathbb{C}^{n^\nu} = \bigoplus e\mathbb{C}^{n^\nu}$$

where in the sum, e runs over the standard tableaux. [Note: Here e is *not* the identity of the group.]

Example:

Let us consider the reduction of rank 3 tensors for $GL(3, \mathbb{C})$. The standard tableaux are

1) $\boxed{1\ 2\ 3}$, 2) $\begin{array}{cc}\boxed{1}&\boxed{2}\\ \boxed{3}\end{array}$,

3) $\begin{array}{cc}\boxed{1}&\boxed{3}\\ \boxed{2}\end{array}$, 4) $\begin{array}{c}\boxed{1}\\ \boxed{2}\\ \boxed{3}\end{array}$.

Let e_i denote the associated QP. Then

$$(e_1 \; F)^{j_1 j_2 j_3} = F^{j_1 j_2 j_3} + F^{j_1 j_3 j_2} + F^{j_2 j_1 j_3} + F^{j_2 j_3 j_1} + F^{j_3 j_1 j_2} + F^{j_3 j_2 j_1},$$

$$(e_2 \; F)^{j_1 j_2 j_3} = F^{j_1 j_2 j_3} + F^{j_2 j_1 j_3} - F^{j_3 j_2 j_1} - F^{j_2 j_3 j_1},$$

$$(e_3 \; F)^{j_1 j_2 j_3} = F^{j_1 j_2 j_3} + F^{j_3 j_2 j_1} - F^{j_2 j_1 j_3} - F^{j_3 j_1 j_2},$$

$$(e_4 \; F)^{j_1 j_2 j_3} = F^{j_1 j_2 j_3} - F^{j_1 j_3 j_2} - F^{j_2 j_1 j_3} + F^{j_2 j_3 j_1} + F^{j_3 j_1 j_2} - F^{j_3 j_2 j_1}.$$

[Here in computing $(e_2 \; F)^{j_1 j_2 j_3}$, for example, we *first* symmetrize $F^{j_1 j_2 j_3}$ in j_1 and j_2 and *then* antisymmetrize in j_1 and j_3. The result is *not* symmetric in j_1 and j_2.] Since

$$F^{j_1 j_2 j_3} = \frac{1}{6}(e_1 \; F)^{j_1 j_2 j_3} + \frac{1}{3}(e_2 \; F)^{j_1 j_2 j_3} + \frac{1}{3}(e_3 \; F)^{j_1 j_2 j_3} + \frac{1}{6}(e_4 \; F)^{j_1 j_2 j_3},$$

we also have

$$C^{n^\nu} = \bigoplus_i e_i C^{n^\nu}$$

for $n = \nu = 3$.

Remark.

If a standard tableau has more than n rows, then $e \; F \equiv 0$ for that tableau. Such tableaux can be ignored.

To prove this statement, note that if the standard tableau has $m = n + k$ $(k \geq 1)$ rows, then $(e \; F)^{(j)}$ has $n + k$ indices j_i in which it is totally antisymmetric. Also at least two of these indices are equal since each j_i cannot exceed n. Hence the result.

LECTURE 27

i) Dimensions of irreducible subspaces of $GL(n, \mathbb{C})$

To each standard tableau with entry i in a box, we can associate an *index diagram* J where the entry i is replaced by j_i, and an associated tensor component $(e\ F)^{j_1, \cdots j_n}$.

Example:

For

$$\begin{array}{|c|c|} \hline 1 & 3 \\ \hline 2 \\ \cline{1-1} \end{array},$$

the index diagram is

$$J = \begin{array}{|c|c|} \hline j_1 & j_3 \\ \hline j_2 \\ \cline{1-1} \end{array}.$$

As each j_i can take on n values, there are actually many J for each standard tableau. Among these, we have the *standard index diagrams* which fulfill the following properties:

    a) The numbers *do not decrease* along each row from left to right,

    b) The numbers *increase* along each column downwards.

    We can also associate a *standard tensor component* (S.T.C.) to each standard index diagram (S.I.D.).

Example:

For rank 3 tensors of $GL(3, \mathbb{C})$ and the standard tableau

$$\begin{array}{|c|c|}\hline 1 & 3 \\\hline 2 \\\cline{1-1}\end{array},$$

the S.I.D.'s are

$$\begin{array}{|c|c|}\hline j_1 & j_2 \\\hline j_3 \\\cline{1-1}\end{array} = \begin{array}{|c|c|}\hline 1 & 1 \\\hline 2 \\\cline{1-1}\end{array}, \quad \begin{array}{|c|c|}\hline 1 & 1 \\\hline 3 \\\cline{1-1}\end{array}, \quad \begin{array}{|c|c|}\hline 1 & 2 \\\hline 2 \\\cline{1-1}\end{array},$$

$$\begin{array}{|c|c|}\hline 1 & 2 \\\hline 3 \\\cline{1-1}\end{array}, \quad \begin{array}{|c|c|}\hline 1 & 3 \\\hline 2 \\\cline{1-1}\end{array}, \quad \begin{array}{|c|c|}\hline 1 & 3 \\\hline 3 \\\cline{1-1}\end{array},$$

$$\begin{array}{|c|c|}\hline 2 & 2 \\\hline 3 \\\cline{1-1}\end{array}, \quad \begin{array}{|c|c|}\hline 2 & 3 \\\hline 3 \\\cline{1-1}\end{array}.$$

The S.T.C.'s are

$$(e\,F)^{j_1 j_2 j_3} = (e\,F)^{112}, (e\,F)^{113}, (e\,F)^{122}, (e\,F)^{123},$$
$$(e\,F)^{132}, (e\,F)^{133}, (e\,F)^{223}, (e\,F)^{233}.$$

Theorem

The dimension of $e\mathbb{C}^{n^\nu}$ for a given standard tableau is equal to the number of S.I.D.'s.

Example:

Consider $n = \nu = 3$. Then we have the following:

Standard Tableau　　　　　　　　　　　S.I.D.'s

1) $\begin{array}{|c|c|c|}\hline 1 & 2 & 3 \\\hline\end{array}$, $\begin{array}{|c|c|c|}\hline j_1 & j_2 & j_3 \\\hline\end{array} = \begin{array}{|c|c|c|}\hline 1 & 1 & 1 \\\hline\end{array}, \begin{array}{|c|c|c|}\hline 1 & 1 & 2 \\\hline\end{array}, \begin{array}{|c|c|c|}\hline 1 & 1 & 3 \\\hline\end{array},$

$\begin{array}{|c|c|c|}\hline 1 & 2 & 2 \\\hline\end{array}, \begin{array}{|c|c|c|}\hline 1 & 2 & 3 \\\hline\end{array}, \begin{array}{|c|c|c|}\hline 1 & 3 & 3 \\\hline\end{array}, \begin{array}{|c|c|c|}\hline 2 & 2 & 2 \\\hline\end{array},$

$\begin{array}{|c|c|c|}\hline 2 & 2 & 3 \\\hline\end{array}, \begin{array}{|c|c|c|}\hline 2 & 3 & 3 \\\hline\end{array}, \begin{array}{|c|c|c|}\hline 3 & 3 & 3 \\\hline\end{array}.$

2) $\begin{array}{|c|c|}\hline 1 & 2 \\\hline 3 \\\cline{1-1}\end{array}$, $\begin{array}{|c|c|}\hline j_1 & j_2 \\\hline j_3 \\\cline{1-1}\end{array} =$ Index diagrams in the preceding example.

3) $\begin{array}{|c|c|}\hline 1 & 3 \\\hline 2 \\\cline{1-1}\end{array}$, $\begin{array}{|c|c|}\hline j_1 & j_3 \\\hline j_2 \\\cline{1-1}\end{array} =$ Index diagrams in the preceding example.

4)
$$\begin{array}{|c|} \hline 1 \\ \hline 2 \\ \hline 3 \\ \hline \end{array} \quad , \quad \begin{array}{|c|} \hline j_1 \\ \hline j_2 \\ \hline j_3 \\ \hline \end{array} \quad = \quad \begin{array}{|c|} \hline 1 \\ \hline 2 \\ \hline 3 \\ \hline \end{array}$$

The dimensions of $e\mathbb{C}^{n^\nu}$ for 1), 2), 3) and 4) are thus 10, 8, 8 and 1 respectively.

Remarks:

1) Note that the *number* of S.I.D.'s depends only on the Young frame. This is consistent with the fact that different standard tableaux with the same frame give equivalent representations. [See below.]

2) For $n = \nu = 4$, the dimension of $\mathbb{C}^{n^\nu}$ is 27. This is 10+8+8+1 as it should be. We can write

$$3 \otimes 3 \otimes 3 = 10 \oplus 8 \oplus 8 \oplus 1.$$

3) For each standard tableau and an associated index diagram J, we have a tensor component which we can denote by $(e\ F)^J$. For instance if

$$\begin{array}{|c|c|} \hline j_1 & j_3 \\ \hline j_2 & \\ \cline{1-1} \end{array} \quad = \quad \begin{array}{|c|c|} \hline 1 & 1 \\ \hline 2 & \\ \cline{1-1} \end{array} \quad ,$$

then

$$(e\ F)^J = (e\ F)^{j_1 j_2 j_3} = (e\ F)^{121}.$$

It may be shown that if J is any index diagram and J_i are standard index diagrams, then

$$(e\ F)^J = \sum_i \xi_i (e\ F)^{j_i}$$

where ξ_i are independent of F [cf. Boerner's book]. We can proceed as follows to find the matrices of the IRR's: The transformation law for $(e\ F)^{J_i}$ is

$$(e\ F)^{J_i} \rightarrow [a]^\nu_{J_i, J'} (e\ F)^{J'}.$$

We can now eliminate nonstandard $(e\ F)^{J'}$ from the right-hand side using the linear relations. The matrix of the transformation can then be read off. This matrix is $d \times d$ where d is the number of S.I.D.'s.

Theorem

The IRR's which arise from the same frame, but different standard tableaux, are equivalent.

For proof, see Boerner's book.

LECTURE 28

i) Classification of irreducible representations of $GL(n, \mathbb{C})$

Theorem (Weyl)

For $GL(n, \mathbb{C})$, the IRR's which arise from tensor spaces of different rank are inequivalent.

For the elements of the matrix

$$[a]^\nu = \underbrace{a \times a \times \ldots \times a}_{\nu \text{ times}}$$

are homogeneous polynomials of degree ν in a_{ij}:

$$[ta]^\nu = t^\nu [a]^\nu, \quad t \in \mathbb{C}.$$

Therefore, the IRR's from the reduction of the representation $\{[a]^\nu\}$ are homogeneous polynomials of degree ν in a_{ij}. But nonzero homogeneous polynomials of different degree cannot be equal. Q.E.D.

The inequivalent IRR's computed in the preceding fashion are denoted by $[\lambda_1, \lambda_2, \ldots, \lambda_n]$. Here λ_i are integers,

$$\lambda_1 \geq \lambda_2 \geq \ldots \geq \lambda_n \geq 0,$$

and $\nu = \Sigma \lambda_i$ is the rank of the tensor space from which the IRR is constructed.

We could have also constructed IRR's in the preceding fashion by reducing the representations $\{[a^*]^\nu\}$. Such IRR's are labeled by $[\mu_1, \mu_2, \ldots, \mu_n]^*$ where μ_i are integers,

$$\mu_1 \geq \mu_2 \geq \ldots \geq \mu_n \geq 0,$$

and $\sum \mu_i$ denotes the rank of the associated tensor space. We put the associated tensor indices which transform with powers of a^* as subscripts. For instance,

$$F_{j_1 j_2 \cdots j_\nu} \to a^*_{j_1 j'_1} a^*_{j_2 j'_2} \cdots a^*_{j_\nu j'_\nu} F_{j'_1 j'_2 \cdots j'_\nu}.$$

The group $GL(n, \mathbb{C})$ has also one-dimensional IRR's denoted by Δ_p and $\Delta^*_q [p, q = 0, \pm 1, \pm 2, \ldots]$. In Δ_p,

$$a \to (\det a)^p$$

while in Δ^*_q,

$$a \to (\det a^*)^q.$$

Putting all these together, we have the following IRR's for $GL(n, \mathbb{C})$:

$$\Delta_p \otimes \Delta^*_q \otimes [\lambda_1, \lambda_2, \ldots, \lambda_{n-1}] \otimes [\mu_1, \mu_2, \ldots, \mu_{n-1}]^*.$$

The transformation of the tensors is (symbolically) given by

$$F^{(j)}_{(k)} \to (\det a)^p (\det a^*)^q [a]^\nu_{(j)(j')} [a^*]^{\nu'}_{(k)(k')} F^{(j')}_{(k')}.$$

Here ν and ν' are the ranks of the tensor spaces:

$$\nu = \sum_i \lambda_i, \quad \nu' = \sum_i \mu_i.$$

Remark.

We have restricted the index i in λ_i and μ_i to $i \leq n-1$ for the following reason: If for example $\lambda_n = 1$, the tensor has n indices as superscripts in which it is totally antisymmetric:

$$F^{j_1 j_2 \cdots j_n \cdots}_{\cdots} = \varepsilon^{j_1 j_2 \cdots j_n} F^{12 \cdots n \cdots}_{\cdots}.$$

Here $\varepsilon^{j_1 j_2 \cdots j_n}$ is the totally antisymmetric tensor with $\varepsilon^{12 \cdots n} = 1$. Since

$$a_{\mu_1 \mu'_1} a_{\mu_2 \mu'_2} \cdots a_{\mu_n \mu'_n} = \det a \, \varepsilon^{\mu_1 \mu_2 \cdots \mu_n},$$

we see that

$$\Delta_p \otimes \Delta^*_q \otimes [\lambda_1 \cdots \lambda_n] \otimes [\mu_1 \mu_2 \cdots \mu_n]^*$$
$$= \Delta_{p+\lambda_n} \otimes \Delta^*_{q+\mu_n} \otimes [\lambda_1 - \lambda_n, \lambda_2 - \lambda_n, \cdots, \lambda_{n-1} - \lambda_n, 0]$$
$$\otimes [\mu_1 - \mu_n, \mu_2 - \mu_n, \cdots, \mu_{n-1} - \mu_n, 0]^*.$$

We can thus delete λ_n and μ_n.

ii) The group $GL(n, \mathbb{R})$

We can construct IRR's of $GL(n, \mathbb{R})$ by restricting the IRR's of $G(n, \mathbb{C})$. Since $a = a^*$ if $a \in GL(n, \mathbb{R})$, we get the following IRR's of $GL(n, \mathbb{R})$:

$$\Delta_p \otimes [\lambda_1, \ \lambda_2, \ \ldots, \ \lambda_{n-1}].$$

iii) The group $SL(n, \mathbb{C})$

If $a \in SL(n, \mathbb{C})$, det $a = 1$. Thus the finite-dimensional IRR's of $SL(n, \mathbb{C})$ are given by

$$[\lambda_1, \ \lambda_2, \ \ldots, \ \lambda_{n-1}] \otimes [\mu_1, \ \mu_2, \ \ldots, \ \mu_{n-1}]^*.$$

We emphasize that these are not all the IRR's of these groups, but only the finite-dimensional ones.

Example:

The IRR's of $SL(2, \mathbb{C})$ are $[\lambda] \otimes [\mu]^*$. Let us find the dimensions of a few of them.

1) The trivial one-dimensional representation can be denoted by $[0] \otimes [0]^*$.

2) The representation $[1] \otimes [0]^*$ has S.I.D.'s

$$\boxed{1} \, , \quad \boxed{2}$$

for the upper indices and dimension 2.

3) The representation $[0] \otimes [1]^*$ is two-dimensional and is the complex conjugate of 2).

4) The representation $[1] \otimes [1]^*$ is clearly four-dimensional. We will see later that $SL(2, \mathbb{C})$ is the universal covering group of the connected Lorentz group $\mathcal{L}_+^\uparrow$. The representation of $\mathcal{L}_+^\uparrow$ on four vectors is this representation of $SL(2, \mathbb{C})$.

5) The representation of $[2] \otimes [0]^*$ has S.I.D.'s

$$\boxed{1 \mid 1} \, , \quad \boxed{1 \mid 2} \, , \quad \boxed{2 \mid 2}$$

and is three-dimensional.

Remark.

The unitary IRR's of $SL(2, \mathbb{C})$ are all infinite-dimensional except for the trivial one. They are all known. [Cf. Naimark's book].

iv) The group $SL(n, \mathbb{R})$

Since $a = a^*$ if $a \in SL(n, \mathbb{R})$, we see that the finite-dimensional IRR's of $SL(n, \mathbb{R})$ are given by

$$[\lambda_1, \lambda_2, \ldots, \lambda_{n-1}].$$

LECTURE 29

i) Tensor methods for compact Lie groups

We now turn to compact Lie groups. The unitary IRR's (UIRR's) of the groups we find below in fact give all their IRR's (up to equivalence). This is because the defining representation $\{a\}$ is faithful and we are reducing the direct products $\{[a]^\nu\} \otimes \{[a^*]^{\nu'}\}$ to find these IRR's. [Cf. the Peter-Weyl theorem.]

ii) The group $U(n)$

If $a \in U(n)$, det $a = (\det a^*)^{-1}$. In the list of IRR's for $GL(n, \mathbb{C})$, we can thus assume that p, $q \geq 0$. Only positive powers of det a and det a^* are in the representations Δ_p, Δ_q^* if p, $q \geq 0$. So we can consider the list $[\lambda_1, \lambda_2, \cdots, \lambda_n] \otimes [\mu_1, \mu_2, \cdots, \mu_n]^*$ and ignore the representations Δ_p, Δ_q^*.

These representations are not IRR for $U(n)$. Let F be a tensor which is traceless for the contraction of any upper index with any lower index:

$$\sum_\lambda F^{\lambda \cdots}_{\lambda \cdots} = 0, \text{ etc.}$$

It remains traceless after a $U(n)$ transformation since

$$\sum_\lambda a_{\lambda\lambda'} a^*{}_{\lambda\mu'} \cdots F^{\lambda' \cdots}_{\mu' \cdots} = \sum_\lambda F^{\lambda \cdots}_{\lambda \cdots}.$$

Thus we can restrict the tensors to be traceless. It can be shown that as a consequence, we need only consider the frames

$$[\lambda_1, \lambda_2, \cdots, \lambda_\rho] \otimes [\mu_1, \mu_2, \cdots, \mu_\sigma]^*, \quad \rho + \sigma = n.$$

The UIRR's of $U(n)$ are normally denoted by

$$[f_1, f_2, \cdots, f_n], \quad f_1 \geq f_2 \geq \cdots \geq f_n$$

where the positivity condition on f_i has been dropped. If $f_i \geq 0$ for $i \leq \rho$ and $f_i < 0$ for $i \geq \rho + 1$, this is the IRR

$$[f_1, f_2, \cdots, f_\rho] \otimes [|f_n|, |f_{n-1}|, \cdots, |f_{\rho+1}|]^*.$$

It is implemented on traceless tensors.

Examples:

1) Consider the group $U(1) = \{e^{i\theta}\}$. Being Abelian, all its UIRR's are one-dimensional. They are given by

$$\begin{array}{ll}
[0] & e^{i\theta} \to 1, \\
[1] & e^{i\theta} \to e^{i\theta}, \\
[2] & e^{i\theta} \to e^{2i\theta}, \\
[-1] & e^{i\theta} \to e^{-i\theta}, \\
\vdots & \vdots \\
[n] & e^{i\theta} \to e^{ni\theta}, \, n = 0, \, \pm 1, \, \pm 2, \, \cdots.
\end{array}$$

2) Consider the group $U(2)$. Some of its UIRR's are the following:

a) $[0, 0]$,   the trivial representation.

b) $[1, 0]$,   the defining two-dimensional representation.

c) $[0, -1]$,   the complex conjugate of the defining representation.

d) $[1, -1]$,   There are four S.I.D.'s. The tensors are also traceless, $\sum_\lambda F_\lambda^\lambda = 0$. Thus the dimension of the UIRR is three.

e) $[2, 0]$,   There are three S.I.D.'s:

$$\boxed{1 \,|\, 1} \,, \quad \boxed{1 \,|\, 2} \,, \quad \boxed{2 \,|\, 2} \,.$$

There is no trace condition since only upper indices are involved. Thus the dimension of the UIRR is three.

Remark.

With the help of the ε tensor, it is possible to raise or lower indices in a tensor in which there is total antisymmetry. The index antisymmetry insures that this transformation is nonsingular.

Example:

For $U(n)$, consider the tensor $F = \{F^{j_1 j_2 \cdots j_{n-1}}\}$ which is totally anti-symmetric. We can then define the dual tensor G by

$$G_j = \varepsilon_{j j_1 j_2 \cdots j_{n-1}} F^{j_1 j_2 \cdots j_{n-1}}.$$

It transforms according to

$$G_j \to \varepsilon_{j j_1 j_2 \cdots j_{n-1}} a_{j_1 j_1'} a_{j_2 j_2'} \cdots a_{j_{n-1} j_{n-1}'} F^{j_1' j_2' \cdots j_{n-1}'} = (\det a) a_{jj'}^* G_{j'}.$$

[Here we have used the identity

$$\varepsilon_{\rho_1 \rho_2 \cdots \rho_k \rho_{k+1}' \cdots \rho_n'} a_{\rho_{k+1}' \rho_{k+1}} a_{\rho_{k+2}' \rho_{k+2}} \cdots a_{\rho_n' \rho_n}$$

$$= (\det a) a_{\rho_1 \rho_1'}^* a_{\rho_2 \rho_2'}^* \cdots a_{\rho_k \rho_k'}^* \, \varepsilon_{\rho_1' \rho_2' \cdots \rho_k' \rho_{k+1} \cdots \rho_n'}$$

valid for $a \in U(n)$. This identity follows from the formula connecting $\det a$ and ε stated previously.] Thus for $U(n)$,

$$[1,\ 1,\ \cdots,\ 0] = \Delta_1 \otimes [1-1,\ 1-1,\ \cdots,\ 0-1] = \Delta_1 \otimes [0,\ 0,\ \cdots,\ -1].$$

In a similar way, one can show that for $U(n)$,

$$[f_1,\ f_2,\ \cdots,\ f_n] = \Delta_k \otimes [f_1 - k,\ f_2 - k,\ \cdots,\ f_n - k],\quad k = 0,\ \pm 1,\ \pm 2,\ \cdots$$

where of course $\Delta_{-k} = \Delta_k^*$. The proof is accomplished by raising or lowering k groups of indices in which the tensor is antisymmetric with the help of the ε tensor. By choosing $k = f_n$, we see that the UIRR's of $U(n)$ are equally well classified by

$$\Delta_p \otimes [f_1,\ f_2,\ \cdots,\ f_{n-1},\ 0] \equiv \Delta_p \otimes [f_1,\ f_2,\ \cdots,\ f_{n-1}],$$

with

$$f_1 \geq f_2 \geq \cdots \geq f_{n-1} \geq 0,\quad p = 0,\ \pm 1,\ \pm 2,\ \cdots.$$

In this form, the absence of trace conditions makes counting of dimensions easier.

Example:

For $U(2)$,

$$[1,\ -1] = \Delta_1^* \otimes [2,\ 0].$$

The S.I.D.'s for $[2, 0]$ are

$$\boxed{1 \mid 1}\ ,\quad \boxed{1 \mid 2}\ ,\quad \boxed{2 \mid 2}\ .$$

Thus the UIRR is of dimension three.

LECTURE 30

i) The group $SU(n)$

The UIRR's of $SU(n)$ can be found from those of $U(n)$. If $a \in U(n)$, then
$$a = (\det a)^{1/n} b$$
where $b \in SU(n)$. Since $(\det a)^{1/n}$ is represented by a multiple of the identity in any UIRR of $U(n)$, the restriction of a UIRR of $U(n)$ to $SU(n)$ remains irreducible for $SU(n)$. Thus the UIRR's of $SU(n)$ are given by
$$[f_1, f_2, \cdots, f_{n-1}], \quad f_1 \geq f_2 \geq f_{n-1} \geq 0.$$
Note that all the UIRR's $[f_1 + k, f_2 + k, \cdots, f_{n-1} + k]$ of $U(n)$ become the same UIRR $[f_1 - f_n, f_2 - f_n, \cdots, f_{n-1} - f_n]$ of $SU(n)$.

Examples:

1) The group $SU(2)$: Here are a few of the UIRR's of SU(2).

Frame	S.I.D.'s	*Dimension*	*"Spin"*
[0]		1	0
[1]	$\boxed{1}$, $\boxed{2}$	2	1/2
[2]	$\boxed{1\,1}$, $\boxed{1\,2}$, $\boxed{2\,2}$	3	1
[3]	$\boxed{1\,1\,1}$, $\boxed{1\,1\,2}$, $\boxed{1\,2\,2}$, $\boxed{2\,2\,2}$	4	3/2

2) The group $SU(3)$:

The UIRR's are denoted by $[p, q]$ with $p \geq q \geq 0$, p and q being integers. The dimensions of $[1, 0]$, $[1, 1]$ and $[2, 1]$ are easily verified to be 3, 3 and 8. For instance, the S.I.D.'s for $[2, 1]$ are

$$\begin{array}{|c|c|}\hline 1 & 1 \\\hline 2 \\\hline\end{array} \,,\quad \begin{array}{|c|c|}\hline 1 & 1 \\\hline 3 \\\hline\end{array} \,,\quad \begin{array}{|c|c|}\hline 1 & 2 \\\hline 2 \\\hline\end{array} \,,\quad \begin{array}{|c|c|}\hline 1 & 2 \\\hline 3 \\\hline\end{array} \,,$$

$$\begin{array}{|c|c|}\hline 1 & 3 \\\hline 2 \\\hline\end{array} \,,\quad \begin{array}{|c|c|}\hline 1 & 3 \\\hline 3 \\\hline\end{array} \,,\quad \begin{array}{|c|c|}\hline 2 & 2 \\\hline 3 \\\hline\end{array} \,,\quad \begin{array}{|c|c|}\hline 2 & 3 \\\hline 3 \\\hline\end{array} \,.$$

For $SU(3)$, the UIRR $[p, q] \equiv [p, q, 0]$ is equivalent to $[p - q, 0, -q]$, Thus the UIRR's can also be denoted by $[\lambda, 0, -\mu]$ where λ and μ are non-negative integers. Here λ denotes the number of upper components and μ the number of lower components of the traceless tensor. The notation $D(\lambda, \mu)$ for $[\lambda, 0, -\mu]$ occasionally appears in the literature. The dimension of $D(\lambda, \mu)$ is

$$\frac{1}{2}(\lambda + 1)(\mu + 1)(\lambda + \mu + 2).$$

ii) The group $O(n)$

If $a \in O(n)$, det $a = \pm 1$. Thus on restriction to $O(n)$,

$$\Delta_p = \Delta_0 \quad \text{if } p \text{ is even,}$$
$$= \Delta_1 \quad \text{if } p \text{ is odd.}$$

Therefore, the $U(n)$ representations become, on restriction to $O(n)$,

$$\Delta_\nu \otimes [f_1, f_2, \cdots, f_{n-1}], \quad \nu = 0, 1.$$

These representations are *not* irreducible for $O(n)$ because tensors which are traceless for the contraction of any two upper indices are invariant under the $O(n)$ action. For instance,

$$F^{\mu_1 \mu_2} \rightarrow a_{\mu_1 \mu_1'} a_{\mu_2 \mu_2'} F^{\mu_1' \mu_2'}$$

so that

$$\sum_\mu F^{\mu\mu}$$

is invariant. The UIRR's of $O(n)$ are given by $\Delta_\nu \otimes [f_1, f_2, \cdots, f_{n-1}]$ where the tensors are required to be traceless. As a consequence, we can restrict the Young frames to those the sum of the lengths of whose first two columns does not exceed n.

Example:

For $O(3)$, the frame

which corresponds to [2, 2] can be discarded. The tensor associated with this frame becomes zero on imposing the trace condition.

Example:

For the group $O(3)$, we list a few of the UIRR's, their dimensions and their common names.

UIRR	Dimension	Name
$\Delta_0 \otimes [0,\ 0]$	1	Scalar
$\Delta_1 \otimes [0,\ 0]$	1	Pseudoscalar
$\Delta_0 \otimes [1,\ 0]$	3	Vector
$\Delta_1 \otimes [1,\ 0]$	3	Pseudovector
$\Delta_0 \otimes [1,\ 1]$	3	Pseudovector
$\Delta_1 \otimes [1,\ 1]$	3	Vector
$\Delta_0 \otimes [2,\ 0]$	5	Symmetric traceless tensor.

For the last row, the tensors are traceless. The angular momenta of these representations are easy to infer. Note that tensor spaces of different rank can lead to the same IRR for $O(n)$. Thus, in the above table, $\Delta_0 \otimes [1,\ 0] = \Delta_1 \otimes [1,\ 1]$ and $\Delta_1 \otimes [1,\ 0] = \Delta_0 \otimes [1,\ 1]$. The notation $\Delta_\nu \otimes [f_1,\ f_2,\ \cdots,\ f_{n-1}]$ therefore overcounts the $O(n)$ IRR's.

iii) The group $SO(n)$

If $a \in SO(n)$, det $a = 1$. Thus $\Delta_0 = \Delta_1$ on restriction of the $O(n)$ representations to $SO(n)$. It follows that the UIRR's of $SO(n)$ are given by $[f_1,\ f_2,\ \cdots,\ f_{n-1}]$. The associated tensors are of course traceless. The Young frames are therefore restricted as in the $O(n)$ case.

Example:

On restriction to $SO(3)$, the distinction between "scalar" and "pseudoscalar" or "vector" and "pseudovector" disappears.

This completes our (incomplete) survey of the representations of $GL(n, \mathbb{C})$ and some of its subgroups. For further information, see the listed references.

PART 4

THE POINCARÉ GROUP

LECTURE 31

i) The connectivity of $O(3,1)$

We recall that the Poincaré group $\mathcal{P}$ is the set $\{(a, \Lambda)|a \in \mathbb{R}^4, \Lambda \in O(3,1)\}$ with the multiplication law

$$(a_1, \Lambda_1)(a_2, \Lambda_2) = (a_1 + \Lambda_1 a_2, \Lambda_1 \Lambda_2).$$

It is also called the inhomogeneous $O(3,1)$ and denoted by $IO(3,1)$. The metric on space-time invariant under the action of $O(3,1)$ is

$$g = \begin{bmatrix} 1 & & & \\ & -1 & & \\ & & -1 & \\ & & & -1 \end{bmatrix}$$

$$= (g_{\mu\nu}), \quad \mu, \nu = 0, 1, 2, 3.$$

We can use this metric to raise and lower indices, thus for example

$$\Lambda_{\mu\nu} = g_{\mu\rho}\Lambda^\rho{}_\nu.$$

Note also the identity

$$g^{\alpha\beta}\Lambda^\mu{}_\alpha\Lambda^\nu{}_\beta = g^{\mu\nu}, \quad g^{\alpha\beta}\Lambda_\alpha{}^\mu\Lambda_\beta{}^\nu = g^{\mu\nu}$$

for $\Lambda = (\Lambda^\mu{}_\nu) \in O(3,1)$. From this equation, we get the identities

$$(\Lambda^0{}_0)^2 - \sum_i(\Lambda^0{}_i)^2 = 1, \quad (\Lambda_0{}^0)^2 - \sum_i(\Lambda_i{}^0)^2 = 1$$

which are useful later.

Topologically $\mathcal{P}$ is $\mathbb{R}^4 \times O(3,1)$. Since $\mathbb{R}^4$ is connected, the connectivity of $\mathcal{P}$ is governed by that of $O(3,1)$. We now examine the connectivity of $O(3,1)$.

We know that

$$\det \Lambda = 1 \text{ or } -1.$$

Also from

$$\Lambda^0{}_0 = \pm[1 + \sum_{i=1}^{3}(\Lambda^0{}_i)^2]^{1/2}$$

we see that

$$\Lambda^0{}_0 \geq 1$$
$$\text{or } \Lambda^0{}_0 \leq -1.$$

Thus $O(3,1)$ has at least four disconnected pieces as shown below.

Notation	Det Λ	$\Lambda^0{}_0$	Contains
$\mathcal{L}_+^\uparrow$	1	≥ 1	$e = 1$,
$\mathcal{L}_-^\uparrow$	-1	≥ 1	Parity P,
$\mathcal{L}_+^\downarrow$	1	≤ -1	Total inversion $I = -1$,
$\mathcal{L}_-^\downarrow$	-1	≤ -1	Time reversal $T = -P$.

It is clear that

$$\mathcal{L}_-^\uparrow = P\, \mathcal{L}_+^\uparrow,$$
$$\mathcal{L}_+^\downarrow = I\, \mathcal{L}_+^\uparrow = PT\, \mathcal{L}_+^\uparrow,$$
$$\mathcal{L}_-^\downarrow = T\, \mathcal{L}_+^\uparrow.$$

We remark on the following. P and T, unlike I are not Lorentz invariant $\Lambda P\Lambda^{-1} \neq P$, $\Lambda T\Lambda^{-1} \neq T$ for all $\Lambda \in \mathcal{L}_+^\uparrow$. Hence their choice depends on the choice of spatial slices and time direction.

We show below that i) $\mathcal{L}_+^\uparrow$ is a group (called the proper orthochronous Lorentz group), and ii) $\mathcal{L}_+^\uparrow$ is connected. Before we do so, it may be pointed out that due to ii), $O(3,1)$ consists of precisely four disconnected pieces. This is because equations like $\mathcal{L}_-^\uparrow = P\, \mathcal{L}_+^\uparrow$ imply that the connectivity of $\mathcal{L}_{+,-}^{\uparrow,\downarrow}$ are the same [Prove this!]. Note also that from i), we can infer that the following are groups:

Definition	Notation	Name
$\mathcal{L}_+^\uparrow \cup \mathcal{L}_-^\uparrow$	$\mathcal{L}^\uparrow$	Orthochronous Lorentz group,
$\mathcal{L}_+^\uparrow \cup \mathcal{L}_+^\downarrow$	$\mathcal{L}_+$	Proper Lorentz group,
$\mathcal{L}_+^\uparrow \cup \mathcal{L}_-^\downarrow$	$\mathcal{L}_0$	Orthochorous Lorentz group.

Proof that $\mathcal{L}_+^\uparrow$ is a group

We have to show that a) $e \in \mathcal{L}_+^\uparrow$, b) if $\Lambda \in \mathcal{L}_+^\uparrow$, then $\Lambda^{-1} \in \mathcal{L}_+^\uparrow$, c) if Λ_1 and $\Lambda_2 \in \mathcal{L}_+^\uparrow$, then $\Lambda_1\Lambda_2 \in \mathcal{L}_+^\uparrow$.

a) is obvious.

For b), note that
$$(\Lambda^{-1})^\mu{}_\nu = \Lambda_\nu{}^\mu.$$
Therefore $(\Lambda^{-1})^0{}_0 = \Lambda^0{}_0 \geq 1$ and of course $\det \Lambda^{-1} = 1$. Thus b) follows.

For c), it is obvious that $\det \Lambda_1\Lambda_2 = +1$. Also
$$(\Lambda_1\Lambda_2)^0{}_0 = (\Lambda_1)^0{}_0(\Lambda_2)_0{}^0 - \sum_i (\Lambda_1)^0{}_i(\Lambda_2)_i{}^0$$

$$= \left| \left[1 + \sum_i (\Lambda_1)^0{}_i(\Lambda_1)^0{}_i \right] \left[1 + \sum_i (\Lambda_2)_i{}^0(\Lambda_2)_i{}^0 \right] \right|^{1/2}$$
$$- \sum_i (\Lambda_1)^0{}_i(\Lambda_2)_i{}^0$$

$$\geq \left| \left[1 + \sum_i (\Lambda_1)^0{}_i(\Lambda_1)^0{}_i \right] \left[1 + \sum_i (\Lambda_2)_i{}^0(\Lambda_2)_i{}^0 \right] \right|^{1/2}$$
$$- \left| \sum_i (\Lambda_1)^0{}_i(\Lambda_1)^0{}_i \sum_j (\Lambda_2)_j{}^0(\Lambda_2)_j{}^0 \right|^{1/2}$$
$$> 0.$$

Here we have used Schwarz's inequality. Since $\Lambda_1\Lambda_2$ is an element of $O(3,1)$, we can thus conclude that $(\Lambda_1\Lambda_2)^0{}_0 \geq 1$. This concludes the proof of c).

Proof that $\mathcal{L}_+^\uparrow$ is connected

Consider the upper sheet
$$H^+ = \{x \in \mathbb{R}^4 | x^2 = 1, \ x^0 > 0\}, \quad x^2 \equiv x^\mu x_\mu$$
of the hyperboloid
$$H = \{x \in \mathbb{R}^4 | x^2 = 1\}.$$
Under the transformation
$$x \to \Lambda x, \quad \Lambda \in \mathcal{L}_+^\uparrow,$$
H^+ is mapped onto itself. For since Λ is a Lorentz transformation, $\Lambda x \in H$. Also
$$(\Lambda x)^0 = \Lambda^0{}_0 x^0 + \Lambda^0{}_i x^i$$

$$= \left[1 + \sum_i (\Lambda^0{}_i)(\Lambda^0{}_i) \right]^{1/2} \left[1 + \sum_i (x^i)^2 \right]^{1/2} + \Lambda^0{}_i x^i$$

which is positive by Schwarz's inequality. Thus $\Lambda x \in H^+$.

Let

$$\hat{x} = (1, \ \vec{0}) \in H^+.$$

For any $x \in H^+$, let L_x be the pure Lorentz transformation in the direction $\vec{x} = (x_1, \ x_2, \ x_3)$ which maps $\hat{x}$ to x:

$$x = L_x \hat{x}.$$

If for instance $\vec{x}$ is in the first direction, L_x is of the form

$$\begin{bmatrix} \cosh\theta & \sinh\theta & 0 & 0 \\ \sinh\theta & \cosh\theta & 0 & 0 \\ 0 & 0 & 1 & 0 \\ 0 & 0 & 0 & 1 \end{bmatrix}.$$

It is easy to see that for each x, L_x is unique. Also, $L_x \in \mathcal{L}_+^\uparrow$. [Gel'fand, Minlos and Shapiro call such an L_x a hyperbolic screw.]

For $\Lambda \in \mathcal{L}_+^\uparrow$, define $R \in \mathcal{L}_+^\uparrow$ by

$$R = L_{\Lambda\hat{x}}^{-1}\Lambda.$$

Since

$$R\hat{x} = L_{\Lambda\hat{x}}^{-1}\Lambda\hat{x} = \hat{x},$$

R is a spatial rotation:

$$R \in SO(3).$$

[We denote the subgroup of spatial rotations in $\mathcal{L}_+^\uparrow$ by $SO(3)$.] Thus given $\Lambda \in \mathcal{L}_+^\uparrow$, we can uniquely determine a point $\Lambda\hat{x}$ in H^+ and an element R of $SO(3)$. That is, we have constructed a map

$$\mathcal{L}_+^\uparrow \xrightarrow{\ \phi\ } H^+ \times SO(3)$$

where

$$\Lambda \xrightarrow{\ \phi\ } (\Lambda\hat{x}, \ L_{\Lambda\hat{x}}^{-1}\Lambda),$$

ϕ is one-to-one. For if Λ_1 and Λ_2 have the same image,

$$\Lambda_1\hat{x} = \Lambda_2\hat{x},$$
$$L_{\Lambda_1\hat{x}}^{-1}\Lambda_1 = L_{\Lambda_2\hat{x}}^{-1}\Lambda_2.$$

From the first equation, we have $L_{\Lambda_1\hat{x}} = L_{\Lambda_2\hat{x}}$, and hence from the second equation, we have $\Lambda_1 = \Lambda_2$. The map is also onto $H^+ \times SO(3)$. For if

$(x, s) \in H^+ \times SO(3)$, the point $L_x s$ in $\mathcal{L}_+^\uparrow$ maps under ϕ to (x, s). Further it is intuitively plausible that ϕ and ϕ^{-1} are continuous.

Thus ϕ is one-to-one, onto and bicontinuous. Because of these nice properties of ϕ, we can conclude that $\mathcal{L}_+^\uparrow$ is topologically the same as $H^+ \times SO(3)$. But since H^+ is topologically like $\mathbb{R}^3$, we have the (topological) identification

$$\mathcal{L}_+^\uparrow = \mathbb{R}^3 \times SO(3).$$

Since $\mathbb{R}^3 \times SO(3)$ is connected, it follows that $\mathcal{L}_+^\uparrow$ is connected. Further since $\mathbb{R}^3$ is simply connected and $SO(3)$ is doubly connected (as we have seen in Lecture 21), $\mathcal{L}_+^\uparrow$ is doubly connected.

We have thus shown that each of the four pieces of $O(3, 1)$ we have listed before is connected and doubly connected.

LECTURE 32

i) The universal covering group of $\mathcal{L}_+^\uparrow$

Since $\mathcal{L}_+^\uparrow$ is doubly connected, its UCG is a simply connected group which covers it twice. This UCG is $SL(2,\mathbb{C})$. The $2-1$ homomorphism $SL(2,\mathbb{C}) \to \mathcal{L}_+^\uparrow$ can be constructed as discussed below. The construction is similar to the construction of the $2-1$ homomorphism $SU(2) \to SO(3)$ explained earlier.

Let τ_i ($i = 1,\ 2,\ 3$) be the Pauli matrices and τ_0 the unit matrix. Let $\tilde{\tau}_0 \equiv \tau_0$ and $\tilde{\tau}_i \equiv -\tau_i$. Then

$$\mathrm{Tr}\,\tilde{\tau}_\mu \tau_\nu = 2g_{\mu\nu}.$$

Any 2×2 hermitian matrix $\hat{M}$ can be written in the form

$$\hat{M} = M^\mu \tau_\mu$$

where M^μ is real and is given by

$$M_\mu = \frac{1}{2}\mathrm{Tr}\,\tilde{\tau}_\mu \hat{M}.$$

Conversely, to every real four-vector x, we can associate the 2×2 hermitian matrix

$$\hat{x} = x^\mu \tau_\mu$$

such that

$$x_\mu = \frac{1}{2}\mathrm{Tr}\,\tilde{\tau}_\mu \hat{x}.$$

Further

$$\det \hat{x} = x^\mu x_\mu.$$

163

If $g \in SL(2, \mathbb{C})$ and $x \in \mathbb{R}^4$, $g\hat{x}g^\dagger$ is hermitian. Hence we can write

$$g\hat{x}g^\dagger = [\Lambda(g)x]^\mu \tau_\mu$$

where $\Lambda(g)$ is a real 4×4 matrix. Since $\det g = 1$,

$$\det [g\hat{x}g^\dagger] = \det \hat{x}.$$

Thus $\Lambda(g) \in O(3, 1)$.

It is easy to verify that the map $g \to \Lambda(g)$ is a $2-1$ homomorphism with kernel ± 1. It may also be shown that the map is onto $\mathcal{L}_+^\uparrow$. Thus $SL(2, \mathbb{C})$ is the universal covering group of $\mathcal{L}_+^\uparrow$.

Note that $\Lambda(g)$ is defined by

$$g\tau_\mu g^\dagger = \tau_\nu \Lambda(g)^\nu{}_\mu.$$

ii) The universal covering group of $\mathcal{P}_+^\uparrow$

The Poincaré group $\mathcal{P}$ consists of four disconnected pieces corresponding to the four pieces of $O(3, 1)$. The component $\mathcal{P}_+^\uparrow$ of $\mathcal{P}$ is $\{(a, \Lambda)\}$ where $a \in \mathbb{R}^4$ and $\Lambda \in \mathcal{L}_+^\uparrow$. The universal covering group $\bar{\mathcal{P}}_+^\uparrow$ of $\mathcal{P}_+^\uparrow$ is $\{(a, \Lambda)\}$ where $a \in \mathbb{R}^4$ and $g \in SL(2, \mathbb{C})$. The multiplication law for $\bar{\mathcal{P}}_+^\uparrow$ is

$$(a_1, g_1)(a_2, g_2) = (a_1 + \Lambda(g_1)a_2, g_1 g_2).$$

iii) The Lie algebra of $\bar{\mathcal{P}}_+^\uparrow$

From general theorems, we know that the Lie algebras of $\bar{\mathcal{P}}_+^\uparrow$ and $\mathcal{P}_+^\uparrow$ are isomorphic. We construct the Lie algebra of $\mathcal{P}_+^\uparrow$.

Since $\{\Lambda\}$ is the pseudo-orthogonal group in its defining representation, when Λ is near the identity, we can write

$$\Lambda = 1 + \frac{i}{2}\varepsilon^{\mu\nu} m_{\mu\nu} + \cdots$$

$$m_{\mu\nu} = -m_{\nu\mu}$$

$$[m_{\mu\nu}]^{\alpha\beta} = -i[\delta_\mu{}^\alpha \delta_\nu{}^\beta - \delta_\mu{}^\beta \delta_\nu{}^\alpha].$$

Let $\{U(a, \Lambda)\}$ be any faithful unitary representation of $\mathcal{P}_+^\uparrow$. If a is small and Λ is near the identity, we can write

$$U(a, \Lambda) = 1 + ia^\mu P_\mu + \frac{i}{2}\varepsilon^{\mu\nu} M_{\mu\nu} + \cdots.$$

We have to find the commutation relations (C.R.'s) of P_μ and $M_{\mu\nu}$ to specify the Lie algebra of $\mathcal{P}_+^\uparrow$.

It is evident that $M_{\mu\nu}$ must fulfill the C.R.'s appropriate to the pseudo-orthogonal group:

$$[M_{\mu\nu},\ M_{\lambda\rho}] = i[g_{\mu\lambda}M_{\nu\rho} + g_{\nu\rho}M_{\mu\lambda} - g_{\mu\rho}M_{\nu\lambda} - g_{\nu\lambda}M_{\mu\rho}].$$

The translations commute and therefore

$$[P_\mu,\ P_\nu] = 0.$$

To find $[M_{\lambda\rho}, P_\mu]$, we start from

$$U(0,\ \Lambda)U(a,\ 1)U(0,\ \Lambda)^{-1} = U(\Lambda a,\ 1)$$

which follows from the group multiplication laws. Inserting

$$U(a,1) = 1 + ia^\mu P_\mu + \cdots,$$

we find, from linear terms in a,

$$U(0,\ \Lambda)P_\mu U(0,\ \Lambda)^{-1} = P_\nu \Lambda^\nu{}_\mu.$$

For $\Lambda = 1 + \frac{i}{2}\varepsilon^{\mu\nu}m_{\mu\nu} + O(\varepsilon^2)$, this gives the required commutator:

$$[M_{\lambda\rho},\ P_\mu] = i[g_{\mu\lambda}P_\rho - g_{\rho\mu}P_\lambda].$$

The Poincaré group has the two Casimir invariants

$$P^2,\ W^2$$

where

$$W_\mu = \frac{1}{2}\varepsilon_{\mu\nu\lambda\rho}P^\nu M^{\lambda\rho}$$

is called the Pauli-Lubanski operator. It is evident that P^2, which is both Lorentz and translation invariant, commutes with P_μ and $M_{\lambda\rho}$. Lorentz invariance of W^2 means that it commutes with $M_{\lambda\rho}$. That it commutes with P_μ as well follows from a simple calculation.

The value of P^2 in a UIRR is the square of mass and the value of W^2/P^2 is the square of spin for a massive particle. We can make the latter statement plausible by noting that in the rest system of a massive particle, W^2/P^2 reduces to $\frac{1}{2}\sum_{i,j=1}^{3} M^{ij}M_{ij}$, which is just the square of the angular momentum associated with the spatial rotation group.

LECTURE 33

i) Galilei group

The Lie algebra of $\mathcal{P}_+^\uparrow$ can be summarized as

$$[J_i, J_j] = i\epsilon_{ijk}J_k, \quad [J_i, K_j] = i\epsilon_{ijk}K_k, \quad [K_i, K_j] = -i\epsilon_{ijk}J_k,$$

$$[J_i, P_j] = i\epsilon_{ijk}P_k, \quad [K_i, P_j] = -i\delta_{ij}P_0, \quad [J_i, P_0] = 0,$$

$$[K_i, P_o] = -iP_i, \quad [P_\mu, P_\nu] = 0.$$

Here $J_i = (1/2)\epsilon_{ijk}M_{jk}$ and $K_i = M_{0i}$. When these generators act on coordinate functions, they are represented by [cf. Lecture 17]

$$J_i = -\frac{i}{2}\epsilon_{ijk}(x_j\frac{\partial}{\partial x_k} - x_k\frac{\partial}{\partial x_j})$$

$$K_i = i(x^0\frac{\partial}{\partial x_i} + x_i\frac{\partial}{\partial x^0}),$$

$$P_\mu = -i\frac{\partial}{\partial x^\mu}.$$

We know that Galilean transformations are obtained from Poincaré transformations by taking the $c \to \infty$ limit where c is the speed of light. To show this, we modify the generators K_i and P_0 as

$$\tilde{K}_i = \frac{K_i}{c} = i(t\frac{\partial}{\partial x_i} + \frac{x_i}{c^2}\frac{\partial}{\partial t}),$$

$$\tilde{P}_0 = cP_0 = -i\frac{\partial}{\partial t},$$

so that the commutation relations become

$$[J_i, J_j] = i\epsilon_{ijk}J_k, \quad [J_i, \tilde{K}_j] = i\epsilon_{ijk}\tilde{K}_k, \quad [\tilde{K}_i, \tilde{K}_j] = -\frac{i}{c^2}\epsilon_{ijk}J_k,$$

$$[J_i, P_j] = i\epsilon_{ijk}P_k, \quad [\tilde{K}_i, P_j] = -\frac{i}{c^2}\delta_{ij}\tilde{P}_0, \quad [J_i, \tilde{P}_0] = 0,$$

$$[\tilde{K}_i, \tilde{P}_o] = -iP_i, \quad [\tilde{P}_0, P_i] = 0.$$

Taking the $c \to \infty$ limit, we obtain

$$[J_i, J_j] = i\epsilon_{ijk}J_k, \quad [J_i, \tilde{K}_j] = i\epsilon_{ijk}\tilde{K}_k, \quad [\tilde{K}_i, \tilde{K}_j] = 0,$$

$$[J_i, P_j] = i\epsilon_{ijk}P_k, \quad [\tilde{K}_i, P_j] = 0, \quad [J_i, \tilde{P}_0] = 0,$$

$$[\tilde{K}_i, \tilde{P}_o] = -iP_i, \quad [\tilde{P}_0, P_i] = 0$$

with

$$\tilde{K}_i = it\frac{\partial}{\partial x_i},$$

$$\tilde{P}_0 = cP_0 = -i\frac{\partial}{\partial t}.$$

This new Lie algebra forms the Galilei group. For instance,

$$e^{i\vec{v}\cdot\vec{K}}\vec{x} = \vec{x} - \vec{v}t,$$

and two consecutive boosts commute. What we have shown here is an example of group contraction.

ii) Group contraction

In order to learn more about group contractions, we consider a Lie algebra L and its subset L'. Remember that L is a vector space. We choose the subset L' to be a vector space too. The basis of L' is $\{e_i, e_j, \cdots\}$ while that of L is $\{e_i, e_j, \cdots, e_\alpha, e_\beta, \cdots\}$. The commutation relations have the following form

$$[e_i, e_j] = C_{ij}^k e_k + C_{ij}^\alpha e_\alpha,$$

$$[e_i, e_\alpha] = C_{i\alpha}^j e_j + C_{i\alpha}^\beta e_\beta,$$

$$[e_\alpha, e_\beta] = C_{\alpha\beta}^i e_i + C_{\alpha\beta}^\gamma e_\gamma.$$

Now we introduce a new basis $\{e_i, e_j, \cdots, \tilde{e}_\alpha, \tilde{e}_\beta, \cdots\}$ where the generators with Greek indices are multiplied by a continuous parameter ϵ:

$$\tilde{e}_\alpha = \epsilon e_\alpha, \tilde{e}_\beta = \epsilon e_\beta, \cdots.$$

The commutation relations of the new generators become

$$[e_i, e_j] = C_{ij}^k e_k + \epsilon^{-1} C_{ij}^\alpha \tilde{e}_\alpha,$$

$$[e_i, \tilde{e}_\alpha] = \epsilon C_{i\alpha}^j e_j + C_{i\alpha}^\beta \tilde{e}_\beta,$$

$$[\tilde{e}_\alpha, \tilde{e}_\beta] = \epsilon^2 C_{\alpha\beta}^i e_i + \epsilon C_{\alpha\beta}^\gamma \tilde{e}_\gamma.$$

When $\epsilon = 1$, we get the original commutation relations. What would happen if we take a singular limit where ϵ approaches zero? We see from above that this limit can be taken if the structure constants of the form C_{ij}^α are all vanishing or equivalently if L' forms a subalgebra. Then, the resulting commutation relations under this limit become

$$[e_i, e_j] = C_{ij}^k e_k,$$

$$[e_i, \tilde{e}_\alpha] = C_{i\alpha}^\beta \tilde{e}_\beta,$$

$$[\tilde{e}_\alpha, \tilde{e}_\beta] = 0.$$

The resulting Lie algebra is said to be contracted from L with respect to L'.

In general, the contraction is achieved in the following way. Let L be a Lie algebra and L' be a subalgebra corresponding to a Lie group G and a subgroup G' respectively. An invertible linear transformation M, depending on some continuous parameters, acts on the algebra L leaving the sub-algebra L' invariant. Now we take a certain limit for the parameters so that the transformation matrix M becomes singular, while the structure constants are kept finite. The resulting Lie algebra is said to be contracted from L with respect to L'. The resulting group is also said to be contracted from G with respect to G'.

In the above example, the transformation matrix $M(\epsilon)$ looks like

$$M(\epsilon) = \begin{bmatrix} I_{(i)} & 0 \\ 0 & \epsilon I_{(\alpha)} \end{bmatrix}$$

and

$$\det(M) = \epsilon^{N(\alpha)},$$

where $N(\alpha)$ denotes the number of generators with Greek indices. Therefore, when ϵ goes to zero this matrix becomes singular.

The transformation matrix for the contraction of the Poincaré group to the Galilei group looks like

$$M(c) = \begin{bmatrix} I_{(3)} & 0 & 0 & 0 \\ 0 & c^{-1}I_{(3)} & 0 & 0 \\ 0 & 0 & c & 0 \\ 0 & 0 & 0 & I_{(3)} \end{bmatrix}$$

acting on the basis $\{\vec{J}, \vec{K}, P^0, \vec{P}\}$ and

$$\det(M) = c^{-2},$$

Therefore, this matrix becomes singular under the limit $c \to \infty$.

iii) Other examples

Example 1. The group $SO(n+1)$ with generators $\{M_{ij} | i, j = 1, 2, \cdots, n+1\}$ has $SO(n)$ as a subgroup:

$$[M_{ij}, M_{kl}] = M_{ik}\delta_{jl} + M_{jl}\delta_{ik} - M_{il}\delta_{jk} - M_{jk}\delta_{il}.$$

Let $N_{ij} = M_{ij}$ with $i, j = 1, 2, \cdots, n$ and $P_i = \epsilon M_{i(n+1)}$. Then N_{ij} satisfies $SO(n)$ commutation relations and

$$[N_{ij}, P_k] = P_i\delta_{jk} - P_j\delta_{ik},$$

$$[P_i, P_j] = -\epsilon^2 N_{ij}.$$

Under the singular limit $\epsilon \to 0$, the contracted algebra is the algebra of the inhomogeneous Euclidean group $\mathcal{E}_n$.

Example 2. The group $SO(n, 1)$ is contracted with respect to the subgroup $SO(n)$ and results in the group $\mathcal{E}_n$. Here, the new generators are defined by $N_{ij} = M_{ij}$ with $i, j = 1, 2, \cdots, n$ and $P_i = \epsilon M_{0i}$.

Example 3. The group $SO(n,1)$ is contracted with respect to the subgroup $SO(n-1,1)$ and results in the group $ISO(n-1,1)$ which is the n-dimensional Poincaré group. The new generators are defined by $N_{ij} = M_{ij}$ with $i,j = 0,1,\cdots,n-1$ and $P_i = \epsilon M_{in}$.

LECTURE 34

i) The UIRR's of $\bar{\mathcal{P}}_+^\uparrow$

Although we will be constructing the UIRR's of $\bar{\mathcal{P}}_+^\uparrow$, we will use the terminology and notation appropriate to $\mathcal{P}_+^\uparrow$. This is inaccurate, but convenient.

Let us introduce some definitions.

If $p \in \mathbb{R}^4$, the *orbit* O_p of p under $\mathcal{L}_+^\uparrow$ is the set of $\mathcal{L}_+^\uparrow p$, that is,

$$O_p = \{\Lambda p | \Lambda \in \mathcal{L}_+^\uparrow\}.$$

The *little*, *stationary* or *isotropy* group G_p of p is the set of all $R \in \mathcal{L}_+^\uparrow$ which leaves p invariant:

$$G_p = \{R | R \in \mathcal{L}_+^\uparrow, \ Rp = p\}.$$

Let $\hat{p}$ be a fixed fiducial point on an orbit O_p. Clearly $O_p = O_{\hat{p}}$. For each $p \in O_{\hat{p}}$, pick *one* $L_p \in \mathcal{L}_+^\uparrow$ such that

$$L_p \hat{p} = p,$$
$$L_{\hat{p}} = 1.$$

We call such a family of L_p as *Wigner boosts* or simply as *boosts*. The hyperbolic screws discussed in Lecture 31 form an example of such a family.

We may note two useful facts here: 1) The little groups G_p and $G_{\Lambda p}$ at any two points on the same orbit are isomorphic. In fact,

$$G_{\Lambda p} = \Lambda G_p \Lambda^{-1}.$$

Proof is trivial. 2) For any $\Lambda \in \mathcal{L}_+^\uparrow$, $L_{\Lambda p}^{-1} \Lambda L_p \equiv R(p, \ \lambda)$ is in the little group of $\hat{p}$. For

$$R(p, \ \Lambda)\hat{p} = L_{\Lambda p}^{-1} \Lambda p = \hat{p}.$$

We call $R(p, \ \Lambda)$ the *Wigner rotation*.

We can construct a UIRR of $\bar{\mathcal{P}}_+^\uparrow$ for each $O_{\hat{p}}$ and each UIRR $\Gamma = \{D(R)\}$ of $G_{\hat{p}}$. The vector space on which this UIRR acts has a basis $|p,\ \lambda\rangle$, where $p \in O_{\hat{p}}$ and the index λ carries the representation Γ as we shall see. We will now specify the action of $U(a,\ \Lambda)$ on $|p,\ \lambda\rangle$. P_μ is diagonal on $|p,\ \lambda\rangle$:

$$P_\mu|p,\ \lambda\rangle = p_\mu|p,\ \lambda\rangle.$$

Hence

$$U(a,\ 1)|p,\ \lambda\rangle = e^{ia^\mu p_\mu}|p,\ \lambda\rangle.$$

Further if $R \in G_{\hat{p}}$,

$$U(0,\ R)|\hat{p},\ \lambda\rangle = |\hat{p},\ \rho\rangle D_{\rho\lambda}(R). \qquad a)$$

Finally,

$$U(0,\ L_p)|\hat{p},\ \lambda\rangle = |p,\ \lambda\rangle. \qquad b)$$

Equations a) and b) specify the action of $U(0,\ \Lambda)$ on $|p,\ \lambda\rangle$. For we can write

$$\Lambda = L_{\Lambda p}R(p,\ \Lambda)L_p^{-1}$$

and therefore

$$\begin{aligned}
U(0,\ \Lambda)|p,\ \lambda\rangle &= U(0,\ L_{\Lambda p})\ U(0,\ R(p,\ \Lambda))|\hat{p},\ \lambda\rangle \\
&= U(0,\ L_{\Lambda p})|\hat{p},\ \rho\rangle D_{\rho\lambda}(R(p,\ \Lambda)) \\
&= |\Lambda p,\ \rho\rangle D_{\rho\lambda}(R(p,\ \Lambda)).
\end{aligned}$$

From this and the relation $U(a,\ \Lambda) = U(a,\ 1)U(0,\ \Lambda)$ follows the action of $U(a,\ \Lambda)$ on $|p,\ \lambda\rangle$:

$$U(a,\ \Lambda)|p,\ \lambda\rangle = e^{ia^\mu(\Lambda p)_\mu}|\Lambda p,\ \rho\rangle D_{\rho\lambda}(R(p,\ \Lambda)).$$

We just mention here without proof that each orbit $O_{\hat{p}}$ and a representation of the little group $G_{\hat{p}}$ define an irreducible representation of $\bar{\mathcal{P}}_+^\uparrow$. In order to learn more about these representations, we now classify the orbits.

ii) The orbits $O_{\hat{p}}$ and the little groups $G_{\hat{p}}$

Figure 1 shows the section of the orbits by the 0-3 plane. The fiducial vectors $\hat{p}$ are shown on each orbit.

I_a: These are orbits associated with massive particles of positive energy. For each $m \neq 0$, we get a different orbit. The little group $G_{\hat{p}}$ is $SU(2)$. Its UIRR describes the spin of the particle.

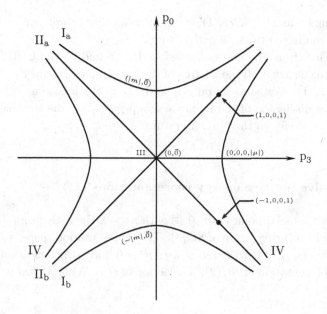

Fig. 1   The section of the orbits by the 0-3 plane

I_b: These are associated with massive particles of negative energy. The little group is $SU(2)$.

$II_{a,b}$: There are only two orbits here. They are associated with massless particles of positive and negative energy respectively. The little groups $G_{\hat{p}}$ for II_a and II_b are isomorphic. For II_a, the Lie algebra of $G_{\hat{p}}$ is spanned by M_{12}, $M_{31} - M_{01}$ and $M_{32} - M_{02}$. For II_b, it is spanned by M_{12}, $M_{31} + M_{01}$ and $M_{32} + M_{02}$. These Lie algebras are isomorphic to the Lie algebra of $\mathcal{E}_2$, the group of Euclidean motions in a plane. However, M_{12} can have half integer eigenvalues in a UIRR, so that the group in question is actually a two-fold covering group $\bar{\mathcal{E}}_2$ of $\mathcal{E}_2$. [cf. Problem 4].

III: The orbit consists of the single vector 0. The little group is $SL(2, \mathbf{C})$.

IV: These are spacelike orbits. For each value of $|\mu| \neq 0$, we get a different orbit. These orbits are associated with tachyons. The little group for any $|\mu|$ is the two-fold covering group $SU(1, 1)$ of $SO(2, 1)$.

Here $SO(2, 1)$ is the subgroup of Lorentz transformations which acts only on p_0, p_1 and p_2. [It is easy to show that $SO(2, 1)$ has two connected components. Here by $SO(2, 1)$, we really mean that component of $SO(2, 1)$ which is connected to the identity.]

It is clear from the above discussion that to find all the UIRR's of $\bar{\mathcal{P}}_+^\uparrow$, we need to construct all the UIRR's of the noncompact groups $\bar{\mathcal{E}}_2$, $SL(2, \mathbf{C})$ and $SU(1, 1)$. We consider only I_a and II_a in the following. For a more exhaustive discussion of the Poincaré group [including the treatment of reflections], we refer to the cited literature.

iii) Massive positive energy representations (I_a)

For a given fiducial point $(m, 0, 0, 0)$ with $m > 0$, the little group is $SU(2)$, the two-fold covering group of $SO(3)$. The representation space is described by a basis $\{|p,\ j\ \lambda\rangle\}$, where $p^2 = m^2$, $p^0 > 0$ and $|j\ \lambda\rangle$ denotes a state of spin j representation of $SU(2)$. The action of $U(a,\ \Lambda)$ on this state is given by

$$U(a,\ \Lambda)|p,\ j\lambda\rangle = e^{ia^\mu(\Lambda p)_\mu}|\Lambda p,\ j\rho\rangle D^j_{\rho\lambda}(R(p,\ \Lambda)),$$

where $D^j_{\rho\lambda}(R(p,\ \Lambda))$ is the spin j representation matrix of $SU(2)$. Therefore, each pair $(m > 0,\ j)$ defines an irreducible representation.

In order to prove the unitarity of this representation, we choose the scalar product as follows:

$$\langle p',\ j\rho|p,\ j\lambda\rangle = 2p_0\delta^3(p' - p)\delta_{\rho\lambda},\quad p,\ p' \in O_{\hat{p}}.$$

The resolution of identity is

$$I = \int d^4p\ \theta(p_0)\ \delta(p^2 - m^2)\sum_\lambda |p,\ j\lambda\rangle\langle p,\ j\lambda|.$$

Since

$$U(a,\ 1)|p,\ j\lambda\rangle\langle p,\ j\lambda|U(a,\ 1)^\dagger = |p,\ j\lambda\rangle\langle p,\ j\lambda|,$$

$U(a,\ 1)$ is unitary:

$$U(a,\ 1)U(a,\ 1)^\dagger = U(a,\ 1)IU(a,\ 1)^\dagger = I.$$

Since

$$U(0,\ R(p,\ \Lambda))\sum_\lambda |p,\ j\lambda\rangle\langle p,\ j\lambda|U^\dagger(0,\ R(p,\ \Lambda)) = \sum_\lambda |p,\ j\lambda\rangle\langle p,\ j\lambda|$$

$[D^j$ being unitary],

$$U(0, \ \Lambda)U(0, \ \Lambda)^\dagger = U(0, \ \Lambda)IU(0, \ \Lambda)^\dagger$$

$$= \int d^4p \ \theta(p_0) \ \delta(p^2 - m^2) \sum_\lambda |\Lambda p, \ j\lambda\rangle\langle\Lambda p, \ j\lambda|$$

which is seen to be I by changing the integration variable from p to $p' = \Lambda p$.
Thus $U(0, \ \Lambda)$ and hence $U(a, \ \Lambda)$ are unitary.

We can choose $L(p)$ to be

$$L(p) = e^{-i\phi M_{12}} e^{-i\theta M_{31}} e^{i\phi M_{12}} e^{-i\delta M_{03}}$$

so that the angles ϕ and θ coincide with the azimuthal angle and the polar
angle of p:

$$L(p)(m, 0, 0, 0) = p$$

where

$$p = (m \cosh \delta, m \sinh \delta \sin \theta \cos \phi, m \sinh \delta \sin \theta \sin \phi, m \sinh \delta \cos \theta).$$

iv) Massless positive energy representations (II_a)

For the orbit II_a, the fiducial point is taken to be $\hat{p} = (\omega, 0, 0, \omega)$. The
Lie algebra of the corresponding little group $G_{\hat{p}}$ is spanned by M_{12}, $M_{31} -
M_{01} = \Pi_1$ and $M_{32} - M_{02} = \Pi_2$. They satisfy

$$[M_{12}, \Pi_1] = i\Pi_2, \quad [M_{12}, \Pi_2] = -i\Pi_1, \quad [\Pi_1, \Pi_2] = 0.$$

The little group $G_{\hat{p}}$ is therefore the two-fold covering group $\bar{\mathcal{E}}_2$ of the Euclidean group $\mathcal{E}_2$ in two dimensions. The vector space of an irreducible
representation of this group is spanned by $|\vec{\xi} \ \eta\rangle$ with $|\vec{\xi}|$ fixed. The two
components of $\vec{\xi}$ denote the eigenvalues of Π_i and η denotes an extra quan-
tum number. There are two cases for η. When $|\vec{\xi}| \neq 0$, $\eta = \pm 1$ and
each sign defines a separate irreducible representation. When $|\vec{\xi}| = 0$,
$\eta = 0, \pm 1/2, \pm 1, \pm 3/2, \cdots$ and each value of η defines a separate irreducible
representation.

The action of $U(\vec{a}, \phi) = e^{i\phi M_{12}} e^{i\vec{a}\cdot\vec{\Pi}}$ on $|\vec{\xi},\ \eta\rangle$ with $|\vec{\xi}| \neq 0$ and $0 \leq \phi < 2\pi$ is given by

$$U(\vec{a}, \phi)|\vec{\xi},\ \eta\rangle = e^{i\vec{a}\cdot\vec{\xi}}|R(\phi)\vec{\xi},\ \eta\rangle,$$

where $R(\phi)\vec{\xi}$ is the vector obtained from $\vec{\xi}$ by the rotation of angle ϕ. Furthermore,

$$U(\vec{a}, \phi + 2\pi)|\vec{\xi},\ \eta\rangle = \eta U(\vec{a}, \phi)|\vec{\xi},\ \eta\rangle.$$

To define a scalar product, we introduce a fixed vector $\hat{\xi}$ and let $\vec{\xi} = R(\phi)\hat{\xi}$. The scalar product we use is

$$\langle \vec{\xi}_1,\ \eta|\vec{\xi}_2,\ \eta\rangle = \delta(\phi_1 - \phi_2).$$

The action of $U(\vec{a}, \phi) = e^{i\phi M_{12}} e^{i\vec{a}\cdot\vec{\Pi}}$ on $|\vec{\xi} = 0,\ \eta\rangle := |\eta\rangle$ is given by

$$U(\vec{a}, \phi)|\eta\rangle = e^{i\phi\eta}|\eta\rangle,$$

with $\langle\eta|\eta\rangle = 1$.

The unitarity of these representations of $\bar{\mathcal{E}}_2$ is obvious and the unitarity of the corresponding representations of the Poincaré group follows. The representation space is spanned by either $\{|\ p, \vec{\xi}, \eta\rangle\}$ for $|\vec{\xi}| \neq 0$ or $\{|\ p, \eta\rangle\}$ for $|\vec{\xi}| = 0$.

The resolution of identity for $|\vec{\xi}| \neq 0$ is

$$I = \int d^4p\ \theta(p_0)\ \delta(p^2) \int_0^{2\pi} d\phi\ |p,\ \vec{\xi},\ \eta\rangle\langle p,\ \vec{\xi},\ \eta|$$

with the scalar product $\langle p', \vec{\xi}', \eta|p, \vec{\xi}, \eta\rangle = 2p_0\delta^3(p' - p)\delta(\phi' - \phi)$.

The resolution of identity for $|\vec{\xi}| = 0$ is

$$I = \int d^4p\ \theta(p_0)\ \delta(p^2)|p,\ \eta\rangle\langle p,\ \eta|$$

with the scalar product $\langle p',\ \eta|p,\ \eta\rangle = 2p_0\delta^3(p' - p)$.

We can fix $L(p)$ as in the massive representation:

$$L(p) = e^{-i\phi M_{12}} e^{-i\theta M_{31}} e^{i\phi M_{12}} e^{-i\delta M_{03}}.$$

Again the angles ϕ and θ coincide with the azimuthal and polar angles of p:

$$L(p)(\omega, 0, 0, \omega) = p$$

where

$$p = (\omega e^\delta, \omega e^\delta \sin\theta \cos\phi, \omega e^\delta \sin\theta \sin\phi, \omega e^\delta \cos\theta).$$

Photons and neutrinos are associated with representations with $\vec{\xi} = 0$. For photons $\eta = \pm 1$ and for neutrinos $\eta = \pm 1/2$.

The spin of a particle is described by the transformation properties of its states under the little group. For a massless particle, this little group is $\bar{\mathcal{E}}_2$ and not $SU(2)$. This is the reason why the spin properties of a massless particle differ from those of a massive particle.

LECTURE 35

i) Reduction of the direct product of two UIRR's

The direct product of two irreducible representation spaces of the Poincaré group $V_1 \otimes V_2$ is irreducible with respect to the direct product of the two Poincaré groups. However, under the diagonal subgroup $\{(\bar{a}, \Lambda) \times (a, \Lambda)\}$ which is isomorphic to the Poincaré group, this space is reducible:

$$V_1 \otimes V_2 = \oplus_n W_n.$$

Here W_n carries an IRR of the (diagonal) Poincaré group.

There are three cases we consider here:

The direct product in question is i) the direct product of two massive irreducible representations $P(m_1, j_1) \otimes P(m_2, j_2)$, ii) the direct product of two massless representations $P(0, \eta_1) \otimes P(0, \eta_2)$ and iii) the direct product of massive and massless representations $P(m, j) \otimes P(0, \eta)$. Here, we restrict ourselves to the positive energy representations and exclude the $|\vec{\xi}| \neq 0$ massless representations. The product states are massive except when two massless states have parallel momenta. We do not consider this exceptional case in this lecture.

ii) Direct product of massive states

We consider a two-particle state with masses, m_1 and m_2, and spins, j_1 and j_2, respectively. A general state can be expressed as a linear sum of the basis states $\{| p_1, j_1\lambda_1 \rangle | p_2, j_2\lambda_2 \rangle \}$ with $p_1^2 = m_1^2$ and $p_2^2 = m_2^2$.

The reduction of the direct product of two massive representations can be summarized by the following formula

181

$$| \lambda_1\lambda_2, \widehat{p}, j\lambda \rangle = \int_{SU(2)} d\mu(g)D^*(g)^j_{\lambda\ \lambda_1-\lambda_2}U(g) \, | \, q_1, j_1\lambda_1 \rangle \, | \, q_2, j_2\lambda_2 \rangle. \quad (*)$$

Here, $d\mu(g)$ is the invariant Haar measure on the $SU(2)$ group manifold which doubly covers $SO(3)$. It is normalized by $\int_{SU(2)} d\mu(g) = 1$. The momenta of two particles are fixed by $q_1 = (q_{10}, 0, 0, q)$ and $q_2 = (q_{20}, 0, 0, -q)$ with positive q_{10}, q_{20} and q. Therefore, the state is described in the center-of-momentum frame and $\widehat{p} = (M, 0, 0, 0)$ with $M = q_{10} + q_{20}$ as the "mass" of the two-particle system.

In order to understand this formula, we first observe that the right-hand side of this formula with $D^*(g)^j_{\lambda\ \lambda_1-\lambda_2}$ substituted with $D^*(g)^j_{\lambda\mu}$, which we simply denote by $| \, \mu\lambda \rangle$, has the following properties:

(1) It transforms like a spin j representation of $SU(2)$.
(2) It is zero unless $\mu = \lambda_1 - \lambda_2$.
(3) The spin j should be equal to or larger than $|\lambda_1 - \lambda_2|$.

Proof of (1)

$$U(g_0) \, | \, \mu\lambda \rangle = \int d\mu(g)D^*(g)^j_{\lambda\mu}U(g_0 g) \, | \, q_1, j_1\lambda_1 \rangle \, | \, q_2, j_2\lambda_2 \rangle$$

$$= \int d\mu(g_0^{-1}\widehat{g})D^*(g_0^{-1}\widehat{g})^j_{\lambda\mu}U(\widehat{g}) \, | \, q_1, j_1\lambda_1 \rangle \, | \, q_2, j_2\lambda_2 \rangle$$

$$= \int d\mu(\widehat{g})D^*(g_0^{-1})^j_{\lambda\rho}D^*(\widehat{g})^j_{\rho\mu}U(\widehat{g}) \, | \, q_1, j_1\lambda_1 \rangle \, | \, q_2, j_2\lambda_2 \rangle$$

$$= D^*(g_0^{-1})^j_{\lambda\rho}\int d\mu(\widehat{g})D^*(\widehat{g})^j_{\rho\mu}U(\widehat{g}) \, | \, q_1, j_1\lambda_1 \rangle \, | \, q_2, j_2\lambda_2 \rangle$$

$$= D^*(g_0^{-1})^j_{\lambda\rho} \, | \, \mu\rho \rangle.$$

Hence $| \, \mu\lambda \rangle$ transforms like a spin j representation of $SU(2)$.

Proof of (2)

We introduce coordinates (α, β, γ) on $SU(2)$ by writing $g \in SU(2)$ as $e^{-i\alpha J_{12}} \, e^{-i\beta J_{31}} \, e^{-i\gamma J_{12}}$. With this coordination, the measure can be written as

$$\int_{SU(2)} d\mu(g) \, \alpha(g) = \frac{1}{16\pi^2} \int_0^{4\pi} d\alpha \int_0^{2\pi} d\gamma \int_{-1}^1 d\cos\beta \, \alpha(g).$$

With $L(p)$ fixed as in the previous lecture,

$$L(p) = e^{-i\phi M_{12}} e^{-i\theta M_{31}} e^{i\phi M_{12}} e^{-i\delta M_{03}},$$

it can be shown that

$$| \ q_1, \ j_1\lambda_1 \rangle = e^{-i\delta^{(1)} M_{03}^{(1)}} | \ \hat{p}_1, \ j_1\lambda_1 \rangle,$$

$$| \ q_2, \ j_2\lambda_2 \rangle = e^{-i\pi M_{31}^{(2)}} e^{-i\delta^{(2)} M_{03}^{(2)}} | \ \hat{p}_2, \ j_2\lambda_2 \rangle,$$

where $m_1 \sinh \delta^{(1)} = m_2 \sinh \delta^{(2)} = q$. Therefore,

$$M_{12}^{(1)} | \ q_1, \ j_1\lambda_1 \rangle = \lambda_1 | \ q_1, \ j_1\lambda_1 \rangle,$$

$$M_{12}^{(2)} | \ q_2, \ j_2\lambda_2 \rangle = -\lambda_2 | \ q_2, \ j_2\lambda_2 \rangle.$$

Here, we have used $e^{-i\pi M_{31}} M_{12} e^{i\pi M_{31}} = -M_{12}$. So

$$U(g) | \ q_1, \ j_1\lambda_1 \rangle | \ q_2, \ j_2\lambda_2 \rangle = e^{-i\alpha M_{12}} e^{-i\beta M_{31}} | \ q_1, \ j_1\lambda_1 \rangle | \ q_2, \ j_2\lambda_2 \rangle$$

with $M_{ij} = M_{ij}^{(1)} + M_{ij}^{(2)}$, and

$$D^*(g)_{\lambda\mu}^j = D^*(e^{-i\alpha M_{12}} e^{-i\beta M_{31}} e^{-i\gamma M_{12}})_{\lambda\mu}^j$$

$$= e^{i\alpha\lambda} D^*(e^{-i\beta M_{31}})_{\lambda\mu}^j e^{i\gamma\mu}$$

$$= e^{i\alpha\lambda} e^{i\gamma\mu} d_{\lambda\mu}^j(\beta).$$

Here $d_{\lambda\mu}^j = D_{\lambda,\mu}^j(e^{-i\beta M_{31}})$ which is real for conventional phase choices [cf. Refs. 31 and 32]. Hence

$$| \ \mu\lambda \rangle = \int d\mu(g) D^*(g)_{\lambda\mu}^j U(g) | \ q_1, \ j_1\lambda_1 \rangle | \ q_2, \ j_2\lambda_2 \rangle$$

$$= \frac{1}{16\pi^2} \int d\alpha \int d\gamma \int d(\cos\beta)$$

$$e^{i\gamma(\mu-\lambda_1+\lambda_2)} e^{i\lambda\alpha} d_{\lambda\mu}^j(\beta) e^{-i\alpha M_{12}} e^{-i\beta M_{31}} | \ q_1, \ j_1\lambda_1 \rangle | \ q_2, \ j_2\lambda_2 \rangle$$

$$= 0$$

due to the γ integration if $\mu \neq \lambda_1 - \lambda_2$. When $\mu = \lambda_1 - \lambda_2$, we denote the state as $| \ \lambda_1\lambda_2, \ \hat{p}, \ j\lambda \rangle$. We have

$$| \ \lambda_1\lambda_2, \ \hat{p}, \ j\lambda \rangle$$

$$= \frac{1}{4\pi} \int_0^{2\pi} d\alpha \int_{-1}^1 d(\cos\beta)$$

$$e^{i\lambda\alpha} d_{\lambda,\lambda_1-\lambda_2}^j(\beta) e^{-i\alpha M_{12}} e^{-i\beta M_{31}} | \ q_1, \ j_1\lambda_1 \rangle | \ q_2, \ j_2\lambda_2 \rangle.$$

Note here that the integration range of α has been reduced to $0 < \alpha < 2\pi$. That is, $\int_0^{4\pi} d\alpha$ has been substituted by $2\int_0^{2\pi} d\alpha$. It is easy to see that the integrand of the right-hand side is invariant under the shift of α by 2π.

To modify the right-hand side further, we consider a two-particle state in the center-of-momentum frame $|\, p_1,\; j_1\lambda_1\rangle \,|\, p_2,\; j_2\lambda_2\rangle_{CM}$ where $p_1 = (p_{10}, \vec{p_1})$ and $p_2 = (p_{20}, \vec{p_2})$ with $\vec{p_1} + \vec{p_2} = 0$. Let (α, β) be the azimuthal and polar angles of $\vec{p_1}$. The corresponding angles of $\vec{p_2}$ are $(\alpha \pm \pi, \pi - \beta)$, where $+$ should be chosen when α is less than π while $-$ should be chosen when α is larger than π. Then,

$$|\, p_1,\; j_1\lambda_1\rangle = L(p_1)\,|\, \hat{p}_1,\; j_1\lambda_1\rangle$$

$$= e^{-i\alpha M_{12}^{(1)}} e^{-i\beta M_{31}^{(1)}} e^{i\alpha M_{12}^{(1)}} e^{-i\delta^{(1)} M_{03}^{(1)}} \,|\, \hat{p}_1,\; j_1\lambda_1\rangle$$

$$= e^{-i\alpha M_{12}^{(1)}} e^{-i\beta M_{31}^{(1)}} e^{-i\delta^{(1)} M_{03}^{(1)}} e^{i\alpha\lambda_1} \,|\, \hat{p}_1,\; j_1\lambda_1\rangle$$

$$= e^{-i\alpha M_{12}^{(1)}} e^{-i\beta M_{31}^{(1)}} e^{i\alpha\lambda_1} \,|\, q_1,\; j_1\lambda_1\rangle.$$

Similarly using $e^{\mp i\pi M_{12}^{(2)}} e^{-i(\pi-\beta) M_{31}^{(2)}} e^{\pm i\pi M_{12}^{(2)}} = e^{i(\pi-\beta) M_{31}^{(2)}}$, we get

$$|\, p_2,\; j_2\lambda_2\rangle = L(p_2)\,|\, \hat{p}_2,\; j_2\lambda_2\rangle$$

$$= e^{-i(\alpha\pm\pi) M_{12}^{(2)}} e^{-i(\pi-\beta) M_{31}^{(2)}} e^{i(\alpha\pm\pi) M_{12}^{(2)}} e^{-i\delta^{(2)} M_{03}^{(2)}} \,|\, \hat{p}_2,\; j_2\lambda_2\rangle$$

$$= e^{-i\alpha M_{12}^{(2)}} e^{-i\beta M_{31}^{(2)}} e^{i\pi M_{31}^{(2)}} e^{-i\delta^{(2)} M_{03}^{(2)}} e^{i\alpha\lambda_2} \,|\, \hat{p}_2,\; j_2\lambda_2\rangle$$

$$= e^{-i\alpha M_{12}^{(2)}} e^{-i\beta M_{31}^{(2)}} e^{i\alpha\lambda_2} \,|\, q_2,\; j_2\lambda_2\rangle.$$

The product state becomes

$$|\, p_1,\; j_1\lambda_1\rangle \,|\, p_2,\; j_2\lambda_2\rangle_{CM}$$
$$= e^{-i\alpha M_{12}} e^{-i\beta M_{31}} e^{i\alpha(\lambda_1+\lambda_2)} \,|\, q_1,\; j_1\lambda_1\rangle \,|\, q_2,\; j_2\lambda_2\rangle. \qquad (**)$$

This shows that

$$e^{-i\alpha M_{12}} e^{-i\beta M_{31}} \,|\, q_1,\; j_1\lambda_1\rangle \,|\, q_2,\; j_2\lambda_2\rangle$$
$$= e^{-i\alpha(\lambda_1+\lambda_2)} \,|\, p_1,\; j_1\lambda_1\rangle \,|\, p_2,\; j_2\lambda_2\rangle_{CM}$$

and we finally get

$$|\, \lambda_1\lambda_2,\; \hat{p},\; j\lambda\rangle$$
$$= \frac{1}{4\pi} \int_0^{2\pi} d\alpha \int_{-1}^{1} d\cos\beta$$
$$d^j_{\lambda,\lambda_1-\lambda_2}(\beta)\, e^{i(\lambda-\lambda_1-\lambda_2)\alpha} \,|\, p_1,\; j_1\lambda_1\rangle \,|\, p_2,\; j_2\lambda_2\rangle_{CM}. \qquad (***)$$

Proof of (3)

The third property is obvious because $D^j_{\lambda\mu} = 0$ if $j < \mu$.

The state in an arbitrary frame can be obtained by a Lorentz transformation as in the single particle case:

$$| \lambda_1\lambda_2, \ p, \ j\lambda \rangle \ = \ U(L(p)) \, | \lambda_1\lambda_2, \ \widehat{p}, \ j\lambda \rangle.$$

The states $| \lambda_1\lambda_2. \ \widehat{p}, \ j\lambda \rangle$ with $\lambda = -j, -j+1, \cdots, j-1, j$ and their Lorentz transformations form a basis of an irreducible representation space labelled by $\{\lambda_1, \lambda_2, M, j\}$. We denote the space as $\tilde{V}(\lambda_1, \lambda_2, M, j)$. We now prove that any state in $V(m_1, j_1) \otimes V(m_2, j_2)$ can be expressed as a linear combination of $| \lambda_1\lambda_2, \ p, \ j\lambda \rangle$ with different $\{\lambda_1, \lambda_2, j, \lambda\}$ so that

$$V(m_1, j_1) \otimes V(m_2, j_2) = \oplus_{\lambda_1, \lambda_2, M, j} \tilde{V}(\lambda_1, \lambda_2, M, j).$$

On the right-hand side of this expression, λ_1 runs from $-j_1$ to j_1 and λ_2 runs from $-j_2$ to j_2. The value of M is lower-bounded by $m_1 + m_2$ and the value of j is lower-bounded by $| \lambda_1 - \lambda_2 |$.

Proof of completeness

Inverting $(*)$, we get

$$U(g) \, | q_1, \ j_1\lambda_1 \rangle \, | q_2, \ j_2\lambda_2 \rangle \ = \ \sum_{j,\lambda} (2j+1) D^j(g)_{\lambda, \lambda_1-\lambda_2} \, | \lambda_1\lambda_2, \widehat{p}, \ j\lambda \rangle.$$

Putting g to be the identity, we obtain

$$| q_1, \ j_1\lambda_1 \rangle \, | q_2, \ j_2\lambda_2 \rangle \ = \ \sum_{j=|\lambda_1-\lambda_2|}^{\infty} (2j+1) \, | \lambda_1\lambda_2, \widehat{p}, \ j \, (\lambda_1-\lambda_2) \rangle. \qquad (****)$$

(This is a formal expression. In order to be more rigorous, we have to consider wave packets.)

Consider an arbitrary product state $| p_1, \ j_1\lambda_1 \rangle \, | p_2, \ j_2\lambda_2 \rangle \in V(m_1, j_1) \otimes V(m_2, j_2)$. Let L be the diagonal boost which maps $\widehat{p}$ to $p_1 + p_2$:

$$L^{-1}(p_1 + p_2) = \widehat{p}.$$

So

$$U(L^{-1}) \, | p_1, \ j_1\lambda_1 \rangle \, | p_2, \ j_2\lambda_2 \rangle = \sum_{\mu_1, \mu_2} C_{\mu_1\mu_2} \, | p_1, \ j_1\mu_1 \rangle \, | p_2, \ j_2\mu_2 \rangle_{CM}$$

which, from $(**)$, can be expressed as

$$\sum_{\mu_1, \mu_2} C_{\mu_1\mu_2} e^{-i\alpha M_{12}} e^{-i\beta M_{31}} e^{i\alpha(\mu_1+\mu_2)} \, | q_1, \ j_1\mu_1 \rangle \, | q_2, \ j_2\mu_2 \rangle$$

$$= \sum_{\mu_1,\mu_2} C'_{\mu_1\mu_2} e^{-i\alpha M_{12}} e^{-i\beta M_{31}} \sum_j (2j+1) \mid \mu_1\mu_2, \widehat{p}, \, j \, (\mu_1 - \mu_2)\rangle$$

where (α, β) are the azimuthal and polar angles of $\vec{p_1}$ and

$$C'_{\mu_1\mu_2} = C_{\mu_1\mu_2} e^{i\alpha(\mu_1+\mu_2)}.$$

Therefore

$$\mid p_1, \, j_1\lambda_1\rangle \mid p_2, \, j_2\lambda_2\rangle = \sum_{\mu_1,\mu_2,j} C''_{\mu_1\mu_2 j} U(L e^{-i\alpha M_{12}} e^{-i\beta M_{31}}) \mid \mu_1\mu_2, \widehat{p}, \, j \, (\mu_1-\mu_2)\rangle$$

with $C''_{\mu_1\mu_2 j} = (2j+1)C'_{\mu_1\mu_2}$. The right-hand side of this equation is just a linear combination of $\{\mid \mu_1\mu_2, p_1 + p_2, \, j\rho\rangle\}$ with different $\{\mu_1, \mu_2, j, \rho\}$, so that the proof is complete.

Another useful relation can be obtained by combining (∗∗) and (∗ ∗ ∗∗):

$$\mid p_1, \, j_1\lambda_1\rangle \mid p_2, \, j_2\lambda_2\rangle_{\text{CM}}$$
$$= \sum_{j,\lambda} (2j+1) e^{-i(\lambda-\lambda_1-\lambda_2)\alpha} d^j{}_{\lambda,\lambda_1-\lambda_2}(\beta) \mid \lambda_1\lambda_2, \, \widehat{p}, \, j\lambda\rangle.$$

This relation and (∗∗∗) tell us how to change the basis between $\{\mid p_1, \, j_1\lambda_1\rangle \mid p_2, \, j_2\lambda_2\rangle_{\text{CM}}\}$ and $\{\mid \lambda_1\lambda_2, \, \widehat{p}, \, j\lambda\rangle\}$. The Clebsch-Gordan coefficients in the center-of-mass frame are

$$\langle k_1, \, j_1\mu_1 \mid \langle k_2, \, j_2\mu_2 \mid \lambda_1\lambda_2, \widehat{p}, \, j\lambda\rangle$$
$$= \frac{k_{10}k_{20}}{\pi \mid \vec{k_1} \mid^2} d^j{}_{\lambda,\lambda_1-\lambda_2}[\beta(\vec{k_1})] e^{i(\lambda-\lambda_1-\lambda_2)\alpha(\vec{k_1})}$$
$$\delta_{\mu_1\lambda_1}\delta_{\mu_2\lambda_2}\delta(\mid \vec{k_1} \mid - \mid \vec{p_1} \mid)\delta^3(\vec{k_1} + \vec{k_2}).$$

The Clebsch-Gordan coefficients in the general frame can be obtained by Lorentz transforming $\mid \lambda_1\lambda_2, \widehat{p}, \, j\lambda\rangle$. We can also calculate the scalar product of these vectors:

$$\langle \mu_1\mu_2, p', \, J\rho \mid \lambda_1\lambda_2, p, \, j\lambda\rangle = \frac{M}{\pi q}\frac{1}{2j+1}\delta_{\mu_1\lambda_1}\delta_{\mu_2\lambda_2}\delta_{Jj}\delta_{\rho\lambda}\delta^4(p'-p),$$

where $M^2 = p^2$ and $q = \sqrt{[M^2 - (m_1+m_2)^2][M^2 - (m_1-m_2)^2]}/2M$.

iii) Direct product of massive and massless states

The reduction of the direct product of massive and massless representations requires the study of

$$\mid \lambda_1\lambda_2, \, \widehat{p}, \, j\lambda\rangle = \int_{SU(2)} d\mu(g) D^*(g)^j{}_{\lambda \, \lambda_1-\lambda_2} U(g) \mid q_1, \, j_1\lambda_1\rangle \mid q_2, \, \lambda_2\rangle.$$

Here, $q_1 = (q_{10}, 0, 0, q)$ and $q_2 = (q, 0, 0, -q)$. Note that the value of λ_2 is fixed while λ_1 runs from $-j_1$ to j_1.

Denoting the irreducible representation space obtained from the above equation as $\tilde{V}(\lambda_1, M, j)$, we get

$$V(m_1, j_1) \otimes V(0, \lambda_2) = \oplus_{\lambda_1, M, j} \tilde{V}(\lambda_1, M, j).$$

On the right-hand side of this expression, λ_1 runs from $-j_1$ to j_1. The value of M is lower-bounded by m_1 and the value of j is lower-bounded by $|\lambda_1 - \lambda_2|$.

Through the same procedure we used in the case of two massive particles, we get

$$| \lambda_1 \lambda_2, \, \widehat{p}, \, j\lambda \rangle = \frac{1}{4\pi} \int_0^{2\pi} d\alpha \int_{-1}^{1} d\cos\beta$$
$$d^j_{\lambda, \lambda_1 - \lambda_2}(\beta) \, e^{i(\lambda - \lambda_1 - \lambda_2)\alpha} \, | p_1, \, j_1\lambda_1 \rangle \, | p_2, \, \lambda_2 \rangle_{\mathrm{CM}}$$

and

$$| p_1, \, j_1\lambda_1 \rangle \, | p_2, \, \lambda_2 \rangle_{\mathrm{CM}}$$
$$= \sum_{j,\lambda} (2j + 1) e^{-i(\lambda - \lambda_1 - \lambda_2)\alpha} d^j_{\lambda, \lambda_1 - \lambda_2}(\beta) \, | \lambda_1 \lambda_2, \, \widehat{p}, \, j\lambda \rangle.$$

From these two equations, we get

$$\langle k_1, \, j_1\mu_1 \, | \, \langle k_2, \, \lambda_2 \, | \, \lambda_1\lambda_2, \widehat{p}, \, j\lambda \rangle = \frac{k_{10}}{\pi|\vec{k_1}|} d^j_{\lambda, \lambda_1 - \lambda_2}[\beta(\vec{k_1})] e^{i(\lambda - \lambda_1 - \lambda_2)\alpha(\vec{k_1})}$$
$$\delta_{\mu_1\lambda_1}\delta(|\vec{k_1}| - |\vec{p_1}|)\delta^3(\vec{k_1} + \vec{k_2}),$$

and

$$\langle \mu_1\lambda_2, p', \, J\rho \, | \, \lambda_1\lambda_2, p, \, j\lambda \rangle = \frac{2M^2}{\pi(M^2 - m_1^2)} \frac{1}{2j + 1} \delta_{\mu_1\lambda_1}\delta_{Jj}\delta_{\rho\lambda}\delta^4(p' - p).$$

iv) Direct product of two massless states

The reduction of the direct product of two massless representations can be formulated likewise by examining

$$| \lambda_1\lambda_2, \, \widehat{p}, \, j\lambda \rangle = \int_{SU(2)} d\mu(g) D^*(g)^j_{\lambda \, \lambda_1 - \lambda_2} U(g) \, | q_1, \, \lambda_1 \rangle \, | q_2, \, \lambda_2 \rangle.$$

Here, $q_1 = (q, 0, 0, q)$ and $q_2 = (q, 0, 0, -q)$. Note that the values of λ_1 and λ_2 are fixed.

Denoting the irreducible representation space obtained from the above equation as $\tilde{V}(M, j)$, we get

$$V(0, \lambda_1) \otimes V(0, \lambda_2) = \oplus_{M,j} \tilde{V}(M, j).$$

On the right-hand side of this expression, M runs over all positive values while the value of j is lower-bounded by $|\lambda_1 - \lambda_2|$. Note here that we neglect the possibility of the exceptional cases where two particles have parallel momenta and $M = 0$.

By the same procedure as in the previous cases, we get

$$| \lambda_1\lambda_2, \, \widehat{p}, \, j\lambda \rangle = \frac{1}{4\pi} \int_0^{2\pi} d\alpha \int_{-1}^{1} d\cos\beta$$
$$d^j{}_{\lambda,\lambda_1-\lambda_2}(\beta) \, e^{i(\lambda-\lambda_1-\lambda_2)\alpha} \, | p_1, \, \lambda_1 \rangle \, | p_2, \, \lambda_2 \rangle_{\text{CM}}$$

and

$$| p_1, \, \lambda_1 \rangle \, | p_2, \, \lambda_2 \rangle_{\text{CM}} = \sum_{j,\lambda}(2j+1)e^{-i(\lambda-\lambda_1-\lambda_2)\alpha}d^j{}_{\lambda,\lambda_1-\lambda_2}(\beta) \, | \lambda_1\lambda_2, \, \widehat{p}, \, j\lambda \rangle.$$

From these two equations, we get

$$\langle k_1, \, \lambda_1 \, | \, \langle k_2, \, \lambda_2 \, | \, \lambda_1\lambda_2, \widehat{p}, \, j\lambda \rangle = \tfrac{1}{\pi}d^j{}_{\lambda,\lambda_1-\lambda_2}[\beta(\vec{k_1})]e^{i(\lambda-\lambda_1-\lambda_2)\alpha(\vec{k_1})}$$

$$\delta(| \vec{k_1} | - | \vec{p_1} |)\delta^3(\vec{k_1} + \vec{k_2}),$$

and

$$\langle \lambda_1\lambda_2, p', \, J\rho \, | \, \lambda_1\lambda_2, p, \, j\lambda \rangle = \frac{2}{\pi(2j+1)}\delta_{Jj}\delta_{\rho\lambda}\delta^4(p' - p).$$

LECTURE 36

i) Direct product of two identical particle states

When we consider two identical particles, the product state must be either symmetrized or anti-symmetrized depending on the spin of the particle. The reduction formula should be modified accordingly.

For the case of massive particles, we get

$$| \lambda_1 \lambda_2 \, \widehat{p} \, J\lambda \rangle_{S,A} = \int_{SU(2)} d\mu(g) D^*(g)^J_{\lambda \text{-} \lambda_1 \text{-} \lambda_2} U(g) \frac{1 \pm \sigma}{2} | q_1 \, j\lambda_1 \rangle \, | q_2 \, j\lambda_2 \rangle.$$

Here, σ is the exchange(flip) operator,

$$\sigma \, | q_1 \, j\lambda_1 \rangle \, | q_2 \, j\lambda_2 \rangle = | q_2 \, j\lambda_2 \rangle \, | q_1 \, j\lambda_1 \rangle$$

and $S(A)$ denotes the symmetric (antisymmetric) state. We take $+$ (symmetric) if the particles are bosons and we take $-$ (antisymmetric) if they are fermions. The diagonal action $U(g)$ of $g \in \bar{\mathcal{P}}^\uparrow_+$ and σ commute and we can write

$$| \lambda_1 \lambda_2 \, \widehat{p} \, J\lambda \rangle_{S,A} = \frac{1 \pm \sigma}{2} | \lambda_1 \lambda_2 \, \widehat{p} \, J\lambda \rangle.$$

The action of σ on $| \lambda_1 \lambda_2 \, \widehat{p} \, J\lambda \rangle$ changes the order of the two one-particle states so that

$$\sigma \, | \lambda_1 \lambda_2 \, \widehat{p} \, J\lambda \rangle = \frac{1}{4\pi} \int_0^{2\pi} d\alpha \int_{-1}^1 d\cos\beta$$
$$d^J{}_{\lambda, \lambda_1 - \lambda_2}(\beta) \, e^{i(\lambda - \lambda_1 - \lambda_2)\alpha} \, | p_2 \, j\lambda_2 \rangle \, | p_1 \, j\lambda_1 \rangle_{\mathrm{CM}}.$$

Here, the momenta of the two particles are given by $p_1 = (p_{10}, \vec{p_1})$ and $p_2 = (p_{20}, -\vec{p_1})$ with the direction of $\vec{p_1}$ denoted by (α, β). Identifying $| p_1 \, j\lambda_1 \rangle$ by $| \overrightarrow{(\alpha, \beta)} \, j\lambda_1 \rangle$, we have

$$| p_2 \, j\lambda_2 \rangle \, | p_1 \, j\lambda_1 \rangle_{\mathrm{CM}} = | \overrightarrow{-(\alpha, \beta)} \, j\lambda_2 \rangle \, | \overrightarrow{(\alpha, \beta)} \, j\lambda_1 \rangle.$$

Using $-\overrightarrow{(\alpha, \beta)} = \overrightarrow{(\alpha + \pi, \pi - \beta)}$, the above state can be written as

$$| p_2\, j\lambda_2\rangle\, | p_1\, j\lambda_1\rangle_{\mathrm{CM}} = |\, \overrightarrow{(\alpha + \pi, \pi - \beta)}\, j\lambda_2\rangle\, |\, -\overrightarrow{(\alpha + \pi, \pi - \beta)}\, j\lambda_1\rangle.$$

We now change the integration variables from α and β to $\tilde\alpha = \alpha + \pi$ and $\tilde\beta = \pi - \beta$ to find

$$\sigma\, |\, \lambda_1\lambda_2\, \widehat{p}\, J\lambda\rangle \;=\; (-1)^{J+\lambda_1+\lambda_2}\tfrac{1}{4\pi}\int_0^{2\pi} d\tilde\alpha \int_{-1}^{1} d\cos\tilde\beta$$

$$d^J{}_{\lambda,\lambda_2-\lambda_1}(\tilde\beta)\, e^{i(\lambda-\lambda_1-\lambda_2)\tilde\alpha}\, |\, \overrightarrow{(\tilde\alpha, \tilde\beta)}\, j\lambda_2\rangle\, |\, -\overrightarrow{(\tilde\alpha, \tilde\beta)}\, j\lambda_1\rangle.$$

Here, we have used the identity:

$$d^J_{\lambda\mu}(\pi - \beta) \;=\; (-1)^{(J+\lambda)} d^J_{\lambda(-\mu)}(\beta).$$

Comparing this with $(***)$ of the previous Lecture, we get

$$\sigma\, |\, \lambda_1\lambda_2\, \widehat{p}\, J\lambda\rangle \;=\; (-1)^{J+\lambda_1+\lambda_2}\, |\, \lambda_2\lambda_1\, \widehat{p}\, J\lambda\rangle,$$

and therefore,

$$|\, \lambda_1\lambda_2\, \widehat{p}\, J\lambda\rangle_{S,A} \;=\; \frac{1}{2}\left(|\, \lambda_1\lambda_2\, \widehat{p}\, J\lambda\rangle \pm (-1)^{(J+\lambda_1+\lambda_2)}\, |\, \lambda_2\lambda_1\, \widehat{p}\, J\lambda\rangle\right).$$

Note here that the two helicities λ_1 and λ_2 may be different. For the case of massless particles, each irreducible representation is characterized by a unique helicity. This is so when we consider only the connected component of the Poincaré group $\mathcal{P}_+^\uparrow$. Massless particle states with different helicities never mix under the Poincaré group $\mathcal{P}_+^\uparrow$. However, the disconnected component of the Poincaré group will mix different helicity states. For example, under parity, helicity changes sign so that the helicity of the photon can be ± 1. Accordingly the reduction formula should be written as

$$V(0; \pm\lambda) \otimes V(0, \pm\lambda) = \bigoplus_{\lambda_1, \lambda_2, M, j} \tilde{V}(M, j).$$

ii) Selection rules and the Landau-Yang theorem

We now consider selection rules for particle decays given by the above analysis. They use *only* the symmetrization postulate and Poincaré invariance. They do not use any quantum field theory(qft).

The relations which determine the selection rules are

$$|\, \lambda_1\lambda_2,\, \widehat{p},\, j\lambda\rangle = \frac{1}{4\pi}\int_0^{2\pi} d\alpha \int_{-1}^{1} d\cos\beta$$

$$d^j{}_{\lambda,\lambda_1-\lambda_2}(\beta)\, e^{i(\lambda-\lambda_1-\lambda_2)\alpha}\, |\, p_1,\, j_1\lambda_1\rangle\, |\, p_2,\, j_2\lambda_2\rangle_{\mathrm{CM}},$$

$$| \lambda_1 \lambda_2, \, \widehat{p}, \, j\lambda \rangle = \frac{1}{4\pi} \int_0^{2\pi} d\alpha \int_{-1}^1 d\cos\beta$$

$$d^j{}_{\lambda, \lambda_1 - \lambda_2}(\beta) \, e^{i(\lambda - \lambda_1 - \lambda_2)\alpha} \, | \, p_1, \, j_1\lambda_1 \rangle \, | \, p_2, \, \lambda_2 \rangle_{\mathrm{CM}},$$

$$| \lambda_1 \lambda_2, \, \widehat{p}, \, j\lambda \rangle = \frac{1}{4\pi} \int_0^{2\pi} d\alpha \int_{-1}^1 d\cos\beta$$

$$d^j{}_{\lambda, \lambda_1 - \lambda_2}(\beta) \, e^{i(\lambda - \lambda_1 - \lambda_2)\alpha} \, | \, p_1, \, \lambda_1 \rangle \, | \, p_2, \, \lambda_2 \rangle_{\mathrm{CM}}$$

and

$$| \lambda_1 \lambda_2 \, \widehat{p} \, J\lambda \rangle_{S,A} = \frac{1}{2} \left(| \, \lambda_1 \lambda_2 \, \widehat{p} \, J\lambda \rangle \pm (-1)^{(J + \lambda_1 + \lambda_2)} \, | \, \lambda_2 \lambda_1 \, \widehat{p} \, J\lambda \rangle \right).$$

Consider the process of decay of a massive particle into two particles which can be either massive or massless. The beginning three relations say that, in the center-of-momentum frame, the coefficient of a possible two-particle final state in the expansion of the initial one-particle state is proportional to $d^j{}_{\lambda, \lambda_1 - \lambda_2}(\beta)$. Therefore, if $d^j{}_{\lambda, \lambda_1 - \lambda_2}(\beta) = 0$, then the corresponding decay is forbidden. For example, a boson cannot decay into one boson and one fermion, for $d^j{}_{\lambda, \lambda_1 - \lambda_2}(\beta) = 0$ when j and λ are integers while $\lambda_1 - \lambda_2$ is an odd half integer. Other forbidden cases are when the spin of the initial particle j is smaller than the difference of helicities of the final two particles $|\lambda_1 - \lambda_2|$. In these cases again $d^j{}_{\lambda, \lambda_1 - \lambda_2}(\beta) = 0$ for an obvious reason.

The above are simple examples. More interesting selection rules come from the last relation. Consider the decay of a massive particle into two identical particles which can be either massive or massless. For a final state with the same helicity $\lambda_1 = \lambda_2 = \lambda$, the last relation becomes

$$| \, \lambda\lambda \, \widehat{p} \, j\lambda \rangle_{S,A} = \frac{1}{2}(1 \pm (-1)^{(j + \lambda_1 + \lambda_2)}) \, | \, \lambda\lambda \, \widehat{p} \, j\lambda \rangle.$$

If $(1 \pm (-1)^{(j + \lambda_1 + \lambda_2)}) = 0$, then

$$| \, \lambda\lambda \, \widehat{p} \, j\lambda \rangle_{S,A} = 0$$

and the decay is forbidden. This fact has nothing to do with $d^j{}_{\lambda, \lambda_1 - \lambda_2}(\beta)$. This selection rule explains the Landau-Yang theorem [C. N. Yang, Phys. Rev. 77, 242 (1950) (The above argument and the generalized selection rules below are from A. P. Balachandran and S. G. Jo, Int. J. Mod. Phys. A22, 6133 (2007))]. It implies that the decay of $Z^0 \to 2\gamma$ is forbidden. The particle Z^0 has spin $j = 1$. Therefore, by considering Z^0 decay at rest we can see that the two photons after decay cannot have opposite helicities. For if the two photons have opposite helicities, then $| \lambda_1 - \lambda_2 | = 2$ and the minimum value for j is 2. This is larger than the spin of Z^0 which is 1. Next we assume that the two photons after decay have the same helicity, i.e. $\lambda_1 = \lambda_2 = \lambda$ which is either 1 or -1. In this case, $d^j{}_{\lambda,\lambda_1-\lambda_2}(\beta) = d^1{}_{\lambda,0}(\beta)$ does not vanish identically. However, the factor $(1 + (-1)^{(j+\lambda_1+\lambda_2)})$ with $j = 1$ and $\lambda_1 = \lambda_2 = \lambda = \pm 1$ vanishes. (Here we use $+$ as the photon is a boson.) This means that the initial state does not contain the corresponding final state and the decay is forbidden.

A more general selection rule can be stated as follows: A massive particle with an odd integer spin cannot decay into two identical particles with the same helicity. This is because

$$1 \pm (-1)^{(j+2\lambda)} = 0,$$

when j is an odd integer regardless of the value of λ. For when λ is an integer, we have to use $+$ and this factor vanishes. When λ is a half-odd integer, we have to use $-$ and again this factor vanishes.

These selection rules require only the Poincaré invariance and the validity of Bose and Fermi statistics. They do not require qft. Their violation will thus have profound consequences for fundamental theories.

PART 5
HOPF ALGEBRAS IN PHYSICS

LECTURE 37

i) On symmetries and symmetry groups

Symmetries arise in physics as a way to formulate equivalences and shared properties between different physical situations. For example, if the Hamiltonian for a system is rotationally invariant, then the energy of a state and the energies of all states we get therefrom by rotations are equal.

There is another aspect of symmetry which is important in physical considerations. If rotations for example are symmetries of a system and we perceive a state of motion of the system such as the earth going around the sun in an elliptical orbit, then we can rotate this orbit with sun as center and get another possible state of motion of the earth. In this way, we can predict more states of motions, and they can be infinite as in the case of rotations.

In quantum theory, this aspect of a symmetry has a more powerful role. Thus even for a finite symmetry group G, if a state vector Ψ changes to a new vector $U(g)\Psi$ under the action of $g \in G$, then we can superpose Ψ and $U(g)\Psi$ with complex coefficients and predict the existence of infinitely many new state vectors. More generally the entire linear manifold spanned by Ψ and its G-transforms are possible state vectors.

Formulation of physical theories is governed by symmetries as well. If observation indicates that the physical world is governed by certain symmetries to a high degree of accuracy, then we can account for these observations by working with Lagrangians and actions exactly or approximately invariant under these symmetries. The Poincaré group is one such symmetry with this role in fundamental theories of nature. Although gauge groups are not symmetries in this sense as we have explained elsewhere,[1,2] they too nowadays provide a metaprinciple controlling the formulation of potential fundamental theories much like the Poincaré group.

Thus symmetries assume many significant roles for a physicist.

Although it has been the practice in physics to rely on groups to define symmetries, we now understand that they provide just certain simple possibilities in this direction and that there exist more general possibilities as well. They are based on Hopf algebras and their variants. Such algebras were first encountered in work on integrable models and conformal field theories and later were given a precise mathematical formulation. [See references 44, 48.] We have now encountered them in fundamental theories as well. [An internet search of the Moyal plane, κ-deformed Poincaré group or doubly special relativity will reveal this.] Thus Hopf algebras and their variants have an emergent importance for physicists. We therefore present an elementary presentation of Hopf algebras in this part.

We first expose the basic features of a group G which let us use it as a symmetry principle. We then formulate them in terms of $\mathbb{C}G$, the group algebra of G, which we discussed in Lecture 10. Now $\mathbb{C}G$ is a Hopf algebra of a special kind, having an underlying group G. We can extract its relevant features to serve as a symmetry and formulate the notion of a general Hopf algebra H. The discussion includes simple examples of H.

ii) Groups as symmetries

Let us assume that we have a quantum theory with a basis for multiparticle state vectors as follows:

$|0\rangle$ = vacuum, $|\rho\rangle$ = single particle state vectors with $\rho \in \{1, 2, \cdots\}$, $|\rho_1\rangle \otimes |\rho_2\rangle$ = two particle state vectors, ... , $|\rho_1\rangle \otimes |\rho_2\rangle \cdots |\rho_N\rangle$ = N-particle state vectors.

Here ρ and ρ_i are for labeling the multiplicity of states. We are treating them as countable for now, ρ, $\rho_i \in \mathbb{Z}$ for convenience, but later this restriction will be discarded. We will have use for these indices below. Also we do not assume that multiparticle vectors change only by a phase when ρ_i is exchanged with ρ_j, $\rho_i \neq \rho_j$. Such symmetrization postulates will be

---

[1]A. P. Balachandran, G. Marmo, B.-S. Skagerstam and A. Stern, *Gauge Symmetries and Fibre Bundles* (Springer-Verlag, Berlin, 1983).

[2]A. P. Balachandran, G. Marmo, B.-S. Skagerstam and A. Stern, *Classical Topology and Quantum States* (World Scientific, Singapore, 1991).

discussed later.

Let G be a symmetry group of the theory with elements g. Then there are unitary representations $U^{(N)} : g \to U^{(N)}(g)$ of G on N-particle states. The action of $U^{(N)}(g)$ on the N-particle states does not change particle number.

The representation U of G on the full Fock space is the direct sum of $U^{(N)}$:

$$U(g) = \bigoplus_N U^{(N)}(g).$$

In the discussion below, we will need the full group algebra $\mathbb{C}G$. We have discussed $\mathbb{C}G$ for finite groups in Lecture 10. Its elements are

$$\alpha^* = \sum_g \alpha(g)g, \quad \beta^* = \sum_h \beta(h)h, \quad g, h \in G$$

whereas their product is

$$\alpha^* \beta^* = \sum_{g,h} \alpha(g)\beta(h)(gh) = \sum_g (\alpha * \beta)(g)g,$$

$\alpha * \beta$ being the convolution product of α and β:

$$(\alpha * \beta)(g) = \sum_h \alpha(gh^{-1})\beta(h).$$

If G is a Lie group, let us assume that it has a left- and right-invariant measure $d\mu$:

$$d\mu(hg) = d\mu(gh) = d\mu(g).$$

Then elements of $\mathbb{C}G$ are

$$\alpha^* = \int d\mu(g)\alpha(g)g, \quad \beta^* = \int d\mu(h)\beta(h)h, \quad g, h \in G$$

and their product is

$$\alpha^* \beta^* = \int d\mu(g)(\alpha * \beta)(g)g,$$

$\alpha * \beta$ being the convolution product of α and β:

$$(\alpha * \beta)(g) = \int d\mu(h)\alpha(gh^{-1})\beta(h).$$

If G is non-compact, appropriate conditions on α and β are to be assumed to ensure the validity of manipulations.

It is important to note for what follows that the representations $U^{(N)}$ and U naturally extend to representations [also called $U^{(N)}$ and U] of $\mathbb{C}G$. They are defined by

$$U^{(N)}(\alpha^*) = \int d\mu(g)\alpha(g)U^{(N)}(g),$$

$$U(\alpha^*) = \int d\mu(g)\alpha(g)U(g).$$

They too preserve particle number.

We now examine $U^{(N)}$. We will see that $U^{(1)}$ determines $U^{(N)}$ $(N \geq 2)$ only if we have a homomorphism Δ from $\mathbb{C}G$ to $\mathbb{C}G \otimes \mathbb{C}G$ (and hence from G into $\mathbb{C}G \otimes \mathbb{C}G$ by restriction):

$$\Delta : g \to \Delta(g),$$

$$\Delta(g_1)\Delta(g_2) = \Delta(g_1 g_2)$$

with certain further properties. The main point is that Δ is not fixed by group theory, but is an independent choice.

1) *The trivial representation $U^{(0)}$*

As the vacuum state is invariant under group action,
$$U^{(0)}(g)|0\rangle = |0\rangle.$$

2) *The one-particle representation $U^{(1)}$*

We can write it as
$$U^{(1)}(g)|\rho\rangle = |\sigma\rangle D_{\rho\sigma}(g)$$
where $D(g)$ is the matrix for $U^{(1)}(g)$.

3) *The two-particle representation $U^{(2)}$*

Group theory does not fix the action of G on two-particle states. We need a homomorphism $\Delta : G \to \mathbb{C}G \otimes \mathbb{C}G$ to fix $U^{(2)}$. Given Δ, we can write
$$U^{(2)}(g) = U^{(1)} \otimes U^{(1)}\Delta(g).$$

A standard choice for Δ is

$$\Delta(g) = g \otimes g.$$

In that case,

$$U^{(2)}(g) = U^{(1)} \otimes U^{(1)} \Delta(g) = U^{(1)}(g) \otimes U^{(1)}(g).$$

But this choice of Δ is not mandatory. It is enough to have a homomorphism Δ (with further properties discussed below) from $\mathbb{C}G$ to $\mathbb{C}G \otimes \mathbb{C}G$. Thus we can set

$$U^{(2)}(g) = U^{(1)} \otimes U^{(1)} \Delta(g)$$

where the representation $U^{(1)}$ of $\mathbb{C}G$ was defined earlier.

We can then extend Δ to a homomorphism from $\mathbb{C}G$ to $\mathbb{C}G \otimes \mathbb{C}G$ by linearity:

$$\Delta\left(\sum \alpha_i g_i\right) = \sum \alpha_i \Delta(g_i), \quad \alpha_i \in \mathbf{C}, \quad g_i \in G.$$

Thus $U^{(2)}$ extends to a homomorphism from $\mathbb{C}G$ to $\mathbb{C}G \otimes \mathbb{C}G$.

4) N-particle representation $U^{(N)}$

Once we are given $\Delta : G \to \mathbb{C}G \otimes \mathbb{C}G$, $U^{(N)}$ is fixed for all $N \geq 2$ by $U^{(1)}$. We have seen how it fixes $U^{(2)}$. We can then set

$$U^{(3)}(g) = \left[U^{(2)} \otimes U^{(1)}\right] \Delta(g)$$

$$U^{(4)}(g) = \left[U^{(3)} \otimes U^{(1)}\right] \Delta(g)$$

etc.

For the choice $\Delta(g) = g \otimes g$, we get the familiar answer

$$U^{(N)}(g) = U^{(1)}(g) \otimes U^{(1)}(g) \otimes \cdots \otimes U^{(1)}(g).$$

Examples

We now consider two examples of Δ which are not just the usual choices. The underlying groups G in these cases are the Euclidean group $\mathcal{E}_3$ and the connected Poincaré group $\mathcal{P}_+^\uparrow$. (In these cases, the natural choices for state vector labels ρ, ρ_i are momenta and components of spin.)

1) $\mathcal{E}_3$

We let

$$\Delta_\theta(g) = F_\theta^{-1}(g \otimes g)F_\theta, \quad g \in \mathcal{E}_3 \tag{1}$$

where

$$F_\theta = e^{\frac{1}{2}P_\mu \theta^{\mu\nu} \otimes P_\nu}$$

with $\theta_{\mu\nu} = -\theta_{\nu\mu} \in \mathbb{R}$ and $P_\mu =$ Translation generators of $\mathcal{E}_3$.

The right-hand side of Eq. (1) is not an element of $G \times G$. Rather it is an element of $\mathbb{C}G \otimes \mathbb{C}G$. Also since it is a similarity transformation of $\Delta_0(g) = g \otimes g$, it is hence evident that Δ_θ is a homomorphism from G to $\mathbb{C}G \otimes \mathbb{C}G$.

2) $\mathcal{P}_+^\uparrow$

This is just a generalization of the $\mathcal{E}_3$ example, with $\mathcal{P}_+^\uparrow$ replacing $\mathcal{E}_3$ and P_μ being translation generators of $\mathcal{P}_+^\uparrow$.

These examples are of particular interest in quantum field theory. They occur in many investigations of quantum physics at the Planck scale, where the appropriate algebra of functions on spacetime is expected to be non-commutative.

Changing the coproduct along the lines of Eq. (1) is called the Drinfel'd twist. [See the references for Hopf algebras.]

LECTURE 38

i) The coassociativity of the coproduct

The standard coproduct Δ_0 on a group G is "coassociative", that is,

$$(\Delta_0 \otimes id)\Delta_0(g) = (id \otimes \Delta_0)\Delta_0(g). \tag{2}$$

Here $\Delta_0(g) = g \otimes g$ and id is the identity map. $\Delta_0 \otimes id$ acts on it by mapping it to $\Delta_0(g) \otimes id(g) = g \otimes g \otimes g$. $id \otimes \Delta_0$ maps $g \otimes g$ to $id(g) \otimes \Delta_0(g) = g \otimes g \otimes g$. So both sides are equal.

Applying $U_1 \otimes U_2 \otimes U_3$ on both sides of Eq. (2), where U_i are representations of G, we see that the transformation law under G of the three-particle state is uniquely fixed because we have coassociativity.

The transformation law of the N-particle is also uniquely fixed when we have coassociativity. That is because we can inductively extend coassociativity to any number of particles. Let us explain how that is so.

The content of Eq. (2) is that applying $\Delta_0 \otimes id$ and $id \otimes \Delta_0$ on $\Delta_0(\mathbb{C}G)$, we end up with the same homomorphism from $\mathbb{C}G$ to $\mathbb{C}G \otimes \mathbb{C}G \otimes \mathbb{C}G$ as the figure below shows.

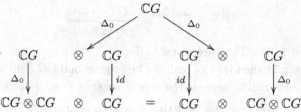

In fact both sides give the same homomorphism

$$g \to g \otimes g \otimes g, \quad g \in G \subset \mathbb{C}G.$$

We can now prove the general result inductively. Suppose that we have

201

deduced a unique homomorphism

$$\mathbb{C}G \to \underbrace{\mathbb{C}G \otimes \mathbb{C}G \otimes \cdots \otimes \mathbb{C}G}_{N \text{ factors}}.$$

Then Eq. (2) says that splitting any two adjacent $\mathbb{C}G$'s on the right-hand side using Δ_0 with the remaining $\mathbb{C}G$'s invariant will give the same homomorphism

$$\mathbb{C}G \to \underbrace{\mathbb{C}G \otimes \mathbb{C}G \otimes \cdots \otimes \mathbb{C}G}_{(N+1) \text{ factors}}.$$

For example

$$\Delta_0 \otimes id \otimes \underbrace{id \otimes \cdots \otimes id}_{(N-2) \text{ factors}}$$

and

$$id \otimes \Delta_0 \otimes \underbrace{id \otimes \cdots \otimes id}_{(N-2) \text{ factors}}$$

split the first two $\mathbb{C}G$. The first is the homomorphism

$$[(\Delta_0 \otimes id)\Delta_0] \otimes \underbrace{id \otimes \cdots \otimes id}_{(N-2) \text{ factors}}$$

from

$$V = \underbrace{\mathbb{C}G \otimes \mathbb{C}G \otimes \cdots \otimes \mathbb{C}G}_{(N-1) \text{ factors}}$$

to

$$W = \underbrace{\mathbb{C}G \otimes \mathbb{C}G \otimes \cdots \otimes \mathbb{C}G}_{(N+1) \text{ factors}}$$

while the second is the homomorphism

$$[(id \otimes \Delta_0)\Delta_0] \otimes \underbrace{id \otimes \cdots \otimes id}_{(N-2) \text{ factors}}$$

also from V to W. They are equal by Eq. (2).

So if we split the first $\mathbb{C}G$ (call it 1) or the second $\mathbb{C}G$ (call it 2), we get the same answer. The same is the case if we split i or $i + 1$. So

$$\text{splitting } 1 = \text{splitting } 2 = \cdots = \text{splitting } N.$$

Thus when Eq. (2) holds, we get a unique homomorphism

$$\mathbb{C}G \to \underbrace{\mathbb{C}G \otimes \mathbb{C}G \otimes \cdots \otimes \mathbb{C}G}_{N \text{ factors}}$$

for all N. This homomorphism is the well-known one

$$g \to g \otimes g \otimes \cdots \otimes g$$

for $g \in G \subset \mathbb{C}G$.

The case of Δ

Consider a general coassociative coproduct Δ:

$$\Delta(g^*)\Delta(h^*) = \Delta(g^*h^*), \quad g^*, h^* \in \mathbb{C}G,$$

$$(\Delta \otimes id)\Delta = (id \otimes \Delta)\Delta.$$

The analysis of Eq. (2) for Δ_0 is adequate to extend it also to Δ. Thus given $\Delta(g^*) \in \mathbb{C}G \otimes \mathbb{C}G$, we can split the first or second $\mathbb{C}G$ using $\Delta \otimes id$ and $id \otimes \Delta$. By induction as above, we can then see that there is a unique homomorphism from $\mathbb{C}G$ to $\mathbb{C}G^{\otimes N}$ for all N.

Applying $U^{(N)} = U^{(1)} \otimes U^{(1)} \otimes \cdots U^{(1)}$ to this unique homomorphism, we can arrive at a unique representation of $\mathbb{C}G$ on N-particle states.

LECTURE 39

i) Example

We want to check the coassociativity of the coproduct
$$\Delta_\theta(g) = F_\theta^{-1}(g \otimes g)F_\theta$$
for the Poincaré group $\mathcal{P}_+^\uparrow$ introduced earlier.

If g is a translation,
$$\Delta_\theta(g) = g \otimes g$$
as translations commute. Thus Δ_θ is coassociative when restricted to translations.

Let us check the coassociativity of Δ_θ when $\mathcal{P}_+^\uparrow$ acts on functions of $\mathbb{R}^4$. If p is four-momentum, $C^\infty(\mathbb{R}^4)$ has the plane wave basis e_p where
$$e_p(x) = e^{ip \cdot x}, \quad p \cdot x = \vec{p} \cdot \vec{x} - p^0 x^0.$$
Translations P_μ act on e_p according to
$$\rho^{(1)}(P\mu)\, e_p = p_\mu e_p$$
where $\rho^{(1)}$ is the representation of $\mathbb{C}\mathcal{P}_+^\uparrow$ on $C^\infty(\mathbb{R}^4)$. Lorentz transformations Λ act on e_p according to
$$\rho^{(1)}(\Lambda)\, e_p = e_{\Lambda p}.$$
The action of $\rho^{(2)}$ of $\mathbb{C}\mathcal{P}_+^\uparrow$ on $e_p \otimes e_q$ is hence
$$\rho^{(2)}(\Lambda)\, e_p \otimes e_q = \left[\rho^{(1)} \otimes \rho^{(1)}\right] \Delta_\theta(\Lambda)\, e_p \otimes e_q.$$

Now $\rho^{(1)} \otimes \rho^{(1)}$ gives a representation of $\mathbb{C}\mathcal{P}_+^\uparrow \otimes \mathbb{C}\mathcal{P}_+^\uparrow$: if $\alpha \otimes \beta \in \mathbb{C}\mathcal{P}_+^\uparrow \otimes \mathbb{C}\mathcal{P}_+^\uparrow$,
$$\rho^{(1)} \otimes \rho^{(1)}(\alpha \otimes \beta) = \rho^{(1)}(\alpha) \otimes \rho^{(1)}(\beta).$$

205

Hence
$$\rho^{(1)} \otimes \rho^{(1)} \Delta_\theta(\Lambda) = \rho^{(1)} \otimes \rho^{(1)}(F_\theta^{-1}) \left[\rho^{(1)}(\Lambda) \otimes \rho^{(1)}(\Lambda) \right] \rho^{(1)} \otimes \rho^{(1)}(F_\theta).$$

Thus
$$\rho^{(2)}(\Lambda) e_p \otimes e_q = \rho^{(1)} \otimes \rho^{(1)}(F_\theta^{-1}) \left[\rho^{(1)}(\Lambda) \otimes \rho^{(1)}(\Lambda) \right] e_p \otimes e_q \, e^{\frac{i}{2} p_\mu \theta^{\mu\nu} q_\nu}$$

$$= \rho^{(1)} \otimes \rho^{(1)}(F_\theta^{-1}) e_{\Lambda p} \otimes e_{\Lambda q} \, e^{\frac{i}{2} p_\mu \theta^{\mu\nu} q_\nu}$$

$$= e_{\Lambda p} \otimes e_{\Lambda q} \, e^{-\frac{i}{2}(\Lambda p)_\mu \theta^{\mu\nu} (\Lambda q)_\nu} \, e^{\frac{i}{2} p_\mu \theta^{\mu\nu} q_\nu}. \tag{3}$$

Let us set
$$E_{p,q} = e_p \otimes e_q.$$
The two-particle state vectors lie in $C^\infty(\mathbb{R}^4) \otimes C^\infty(\mathbb{R}^4) \equiv C^\infty(\mathbb{R}^4)^{\otimes 2}$, that is in the span of $E_{p,q}$. $\rho^{(2)}$ gives an action of $\mathbb{CP}_+^\uparrow$ on $C^\infty(\mathbb{R}^4)^{\otimes 2}$.

A basis for $C^\infty(\mathbb{R}^4)^{\otimes 3}$ is $e_p \otimes e_q \otimes e_r = E_{p,q} \otimes e_r = e_p \otimes E_{q,r}$. We can act on these vectors either with
$$\rho^{(2)} \otimes \rho^{(1)} \Delta_\theta(\Lambda)$$
or with
$$\rho^{(1)} \otimes \rho^{(2)} \Delta_\theta(\Lambda).$$
They are equal if Δ_θ is coassociative. To see that they are in fact equal, let us write out each of these representations:
$$\rho^{(2)} \otimes \rho^{(1)} \Delta_\theta(\Lambda) E_{p,q} \otimes e_r = \rho^{(2)} \otimes \rho^{(1)} F_\theta^{-1} (\Lambda \otimes \Lambda) F_\theta E_{p,q} \otimes e_r.$$
We can evaluate this using Eq. (3) and remembering that $\rho^{(2)}(P_\mu) E_{p,q} = (p+q)_\mu E_{p,q}$. So we set $p+q$ for p and r for q in Eq. (3) to get
$$\rho^{(2)} \otimes \rho^{(1)} F_\theta^{-1} (\Lambda \otimes \Lambda) F_\theta E_{p,q} \otimes e_r$$

$$= \left[\rho^{(2)} \otimes \rho^{(1)} (\Lambda \otimes \Lambda) E_{p,q} \otimes e_r \right] e^{-\frac{i}{2}[\Lambda(p+q)]_\mu \theta^{\mu\nu} (\Lambda r)_\nu} \, e^{\frac{i}{2}(p+q)_\mu \theta^{\mu\nu} r_\nu}.$$

This is, again using Eq. (3),
$$e_{\Lambda p} \otimes e_{\Lambda q} \otimes e_{\Lambda r} \, e^{-\frac{i}{2}(\Lambda p)_\mu \theta^{\mu\nu} (\Lambda q)_\nu} \, e^{\frac{i}{2} p_\mu \theta^{\mu\nu} q_\nu} \, e^{-\frac{i}{2}[\Lambda(p+q)]_\mu \theta^{\mu\nu} (\Lambda r)_\nu} \, e^{\frac{i}{2}(p+q)_\mu \theta^{\mu\nu} r_\nu}. \tag{4}$$

In a similar way we can evaluate
$$\rho^{(1)} \otimes \rho^{(2)} \Delta_\theta(\Lambda) e_p \otimes E_{q,r}$$
to find
$$\rho^{(1)} \otimes \rho^{(2)} \Delta_\theta(\Lambda) e_p \otimes E_{q,r}$$

$$= e_{\Lambda p} \otimes e_{\Lambda q} \otimes e_{\Lambda r}$$

$$e^{-\frac{i}{2}(\Lambda p)_\mu \theta^{\mu\nu} [\Lambda(q+r)]_\nu} \, e^{\frac{i}{2} p_\mu \theta^{\mu\nu} (q+r)_\nu} \, e^{-\frac{i}{2}(\Lambda q)_\mu \theta^{\mu\nu} (\Lambda r)_\nu} \, e^{\frac{i}{2} q_\mu \theta^{\mu\nu} r_\nu}. \tag{5}$$

We see that Eq. (4) and Eq. (5) are equal. Thus we get a unique representation $\rho^{(3)}$ on three-particle states using the coproduct.

LECTURE 40

i) Further properties of $\mathbb{C}G$

The group algebra $\mathbb{C}G$ has further important properties besides the coproduct. They are significant for physics. Let us now discuss them.

1) A group G has the "trivial" one-dimensional representation ϵ:

$$\epsilon : G \to \mathbb{C},$$

$$\epsilon(g) = 1.$$

It extends to $\mathbb{C}G$ to give a one-dimensional representation,

$$\epsilon : \int d\mu(g)\alpha(g)g \to \int d\mu(g)\alpha(g)$$

and is then called a counit.

We need ϵ to define state vectors such as the vacuum and operators invariant by G. Under $\mathbb{C}G$, they transform by the representation ϵ.

2) Every element $g \in G$ has an inverse g^{-1}. We define an antipode S on G as the map $g \to g^{-1}$:

$$S : G \to G$$
$$g \to S(g) = g^{-1}.$$

It is an antihomomorphism:

$$S(gh) = S(h)S(g).$$

It extends by linearity to an antihomomorphism from $\mathbb{C}G$ to $\mathbb{C}G$:

$$S : \hat{\alpha} = \int d\mu(g)\alpha(g)g \to S(\hat{\alpha}) = \int d\mu(g)\alpha(g)S(g),$$
$$S(\hat{\alpha}\hat{\beta}) = S(\hat{\beta})S(\hat{\alpha}).$$

3) There is a compatibility condition fulfilled by multiplication and Δ. It is a generalized version of the requirement that they commute. Thus let m denote the multiplication map:

$$m(g \otimes h) = gh, \quad g, h \in G.$$

It extends by linearity to $\mathbb{C}G$:

$$m \left(\int d\mu(g)\alpha(g)g \otimes \int d\mu(h)\beta(h)h \right)$$

$$= \int d\mu(g)d\mu(h)\alpha(g)\beta(h)gh.$$

Then

$$\Delta \circ m(a \otimes b) = (m_{13} \otimes m_{24})(\Delta \otimes \Delta)(a \otimes b).$$

Here

$$(m_{13} \otimes m_{24})(a_1 \otimes a_2 \otimes a_3 \otimes a_4) = a_1 a_3 \otimes a_2 a_4.$$

Let us check this for $g, h \in G$,

$$\text{Left-hand side} = \Delta(gh) = (gh) \otimes (gh).$$

$$\text{Right-hand side} = m_{13} \otimes m_{24}(g \otimes g \otimes h \otimes h) = (gh) \otimes (gh) = \text{Left-hand side}.$$

4) In quantum theory, we are generally interested in unitary representations U of G, which means that

$$U(g^{-1}) = U(g)^*,$$

$*$ denoting the adjoint operation.

We want to express this equation in terms of a $*$-operation from $\mathbb{C}G$ to $\mathbb{C}G$. It is to be an anti-linear anti-homomorphism just as the adjoint operation:

$$(\lambda\hat{\alpha})^* = \bar{\lambda}\hat{\alpha}^*, (\hat{\alpha}\hat{\beta})^* = \hat{\beta}^*\hat{\alpha}^*, \quad \lambda \in \mathbb{C}, \quad \hat{\alpha}, \hat{\beta}, \hat{\alpha}^*, \hat{\beta}^* \in \mathbb{C}G.$$

Such a $*$ exists for $\mathbb{C}G$:

$$\left(\int d\mu(g)\alpha(g)g \right)^* = \int d\mu(g)\bar{\alpha}(g)g^{-1}.$$

Note that S and $*$ are different operations since for a given complex α,

$$S(\int d\mu(g)\alpha(g)g) = \int d\mu(g)\alpha(g)g^{-1} \neq \int d\mu(g)\bar{\alpha}(g)g^{-1}.$$

The unitarity condition on the representation U now translates to the condition

$$U(\hat{\alpha}^*) = U(\hat{\alpha})^*.$$

U is thus "$*$-representation" of $\mathbb{C}G$.

LECTURE 41

i) Introducing Hopf algebras

The definition of a Hopf algebra H can be formulated by extracting the appropriate properties of $\mathbb{C}G$ so that it can be used as a symmetry algebra in quantum theory. For this reason, it is also called a quantum group.

A $*$-Hopf algebra H is an associative, $*$-algebra with unit element e:

1) $(\alpha\beta)\gamma = \alpha(\beta\gamma)$ for $\alpha, \beta, \gamma \in H$

2) $* : \alpha \in H \to \alpha^* \in H$,
   $(\alpha\beta)^* = \beta^*\alpha^*$, $\alpha, \beta \in H$,
   $(\lambda\alpha)^* = \bar{\lambda}\alpha^*$, $\lambda \in \mathbb{C}$, $\alpha \in H$.

3) $e^* = e$, $\alpha e = e\alpha = \alpha$, $\forall \alpha \in H$.

It is equipped with a coproduct or comultiplication Δ, a counit ϵ and an antipode S with the following properties:

1) The coproduct

$$\Delta : H \to H \otimes H \tag{6}$$

is a $*$-homomorphism.

2) The counit

$$\epsilon : H \to \mathbb{C}$$

is a $*$-homomorphism subject to the conditions

$$(id \otimes \epsilon)\Delta = id = (\epsilon \otimes id)\Delta. \tag{7}$$

3) The antipode

$$S : H \to H$$

is a $*$-antihomomorphism,

$$S(\alpha\beta) = S(\beta)S(\alpha)$$

209

subject to the following conditions. Define

$$m_r(\xi \otimes \eta \otimes \rho) = \eta\xi \otimes \rho,$$

$$m'_r(\xi \otimes \eta \otimes \rho) = \rho \otimes \eta\xi.$$

Then

$$m_r\left[(S \otimes id \otimes id)(id \otimes \Delta)\Delta(\alpha)\right] = e \otimes \alpha,$$

$$m'_r\left[(id \otimes S \otimes id)(id \otimes \Delta)\Delta(\alpha)\right] = \alpha \otimes e, \quad \alpha \in H. \tag{8}$$

4) It is convenient to write multiplication in terms of the multiplication map m:

$$m(\alpha \otimes \beta) = \alpha\beta.$$

Then, as for $\mathbb{C}G$, we can also define $m_{13} \otimes m_{24}$:

$$(m_{13} \otimes m_{24})(\alpha_1 \otimes \alpha_2 \otimes \alpha_3 \otimes \alpha_4) = \alpha_1\alpha_3 \otimes \alpha_2\alpha_4.$$

This operation extends by linearity to all of $H \otimes H \otimes H \otimes H$. Then we require as for $\mathbb{C}G$ that

$$\Delta m(\alpha \otimes \beta) = (m_{13} \otimes m_{24})(\Delta \otimes \Delta)(\alpha \otimes \beta). \tag{9}$$

This is a compatibility condition between m and Δ.

Remarks

a) Some of the above conditions appear more transparent in Kuperberg's diagramatic notation. It will be discussed later.

b) The statement that Δ is a $*$-homomorphism is complete only if a $*$-operation is defined on $H \otimes H$. The following two choices are possible:

$$(\alpha \otimes \beta)^* = \alpha^* \otimes \beta^*,$$

$$(\alpha \otimes \beta)^* = \beta^* \otimes \alpha^*.$$

c) The meaning of the relation 2) can be clarified by verifying it for $\mathbb{C}G$ and the conventional choice $\Delta(g) = g \otimes g, \quad g \in G \subset \mathbb{C}G$ of Δ:

$$(id \otimes \epsilon)\Delta(g) = (id \otimes \epsilon)(g \otimes g) = g \otimes_{\mathbb{C}} 1 = id(g),$$

and hence

$$(id \otimes \epsilon)\Delta \int d\mu(g)\alpha(g)g = \int d\mu(g)\alpha(g)(id \otimes \epsilon)\Delta(g) = id\left(\int d\mu(g)\alpha(g)g\right).$$

Similarly, $(id \otimes \epsilon)\Delta = id$ for $\mathbb{C}G$.

d) We can easily check the validity of Eq. (8) for $\mathbb{C}G$ and the conventional choices of S and Δ.

e) We get a "Hopf algebra" by dropping the statements in the above which involve the $*$.

f) A *bi-algebra* $\mathcal{B}$ has less structure than a Hopf algebra: Thus

-It is an algebra, that is, has the multiplication map m.

-It is a "coalgebra", that is, has the coproduct $\Delta : \mathcal{B} \to \mathcal{B} \otimes \mathcal{B}$ which is a homomorphism.

-m and Δ fulfill the compatibility condition Eq. (9).

There are several known examples of Hopf algebras with applications in physics. One such well-known example is $U_q\left(SL(2, \mathbb{C})\right)$. [See references given under 'Hopf Algebras' at the end.] The Poincaré algebra with the twisted coproduct is another such example. It is a non-trivial example as the twisted coproduct of the Poincaré group element involves functions of momenta which cannot be expressed in terms of group elements.

We now turn to the notion of statistics of identical particles and its role in the theory of Hopf algebras.

LECTURE 42

i) Identical particles and statistics

We begin the discussion with a familiar example. Consider a particle in quantum theory with its associated space V of state vectors. Let a symmetry group G act on V by a representation ρ.

The associated two-particle vector space is based on $V \otimes V$. For the conventional choice of the coproduct Δ,

$$\Delta(g) = g \otimes g, \quad g \in G \subset \mathbb{C}G,$$

the symmetry group G acts on $\xi \otimes \eta \in V \otimes V$ according to

$$\xi \otimes \eta \rightarrow (\rho \otimes \rho)\Delta(g)\xi \otimes \eta = \rho(g)\xi \otimes \rho(g)\eta.$$

Let τ be the "flip operator":

$$\tau(\xi \otimes \eta) = \eta \otimes \xi.$$

It is to be a linear operator so that

$$\tau \sum_i \xi_i \otimes \eta_i = \sum_i \eta_i \otimes \xi_i.$$

Note that τ is <u>not</u> an element of $(\rho \otimes \rho)\Delta(\mathbb{C}G)$:

$$\tau \notin \{(\rho \otimes \rho)\Delta(g^*)|g^* \in \mathbb{C}G\} \equiv (\rho \otimes \rho)\Delta(\mathbb{C}G).$$

Now τ commutes with $(\rho \otimes \rho)\Delta(g)$:

$$(\rho \otimes \rho)\Delta(g)\xi \otimes \eta = \rho(g)\eta \otimes \rho(g)\xi$$

$$= \tau(\rho \otimes \rho)\Delta(g)\xi \otimes \eta.$$

Hence also

$$\tau(\rho \otimes \rho)\Delta(\mathbb{C}G) = (\rho \otimes \rho)\Delta(\mathbb{C}G)\tau.$$

213

Further

$$\tau^2 = 1 \otimes 1$$

so that τ generates the permutation group S_2.

Now since τ commutes with the action of $\mathbb{C}G$ on $V \otimes V$, the eigenspaces of τ for eigenvalues ± 1 are invariant by the action of $\mathbb{C}G$. These eigenspaces are obtained by symmetrization and anti-symmetrization:

$$V \otimes_S V = \frac{1+\tau}{2} V \otimes V = \left\{ \frac{1+\tau}{2} \xi \otimes \eta = \frac{1}{2}(\xi \otimes \eta + \eta \otimes \xi), \xi, \eta \in V \right\}$$

$$V \otimes_A V = \frac{1-\tau}{2} V \otimes V = \left\{ \frac{1-\tau}{2} \xi \otimes \eta = \frac{1}{2}(\xi \otimes \eta - \eta \otimes \xi) \right\}.$$

$V \otimes_S V$ describes bosons and $V \otimes_A V$ describes fermions.

We can easily extend this argument to N-particle states. Thus on $V \otimes V \otimes \cdots \otimes V$, we can define transposition operators

$$\tau_{i,i+1} = \underbrace{1 \otimes \cdots \otimes 1}_{(i-1) \text{ factors}} \otimes \tau \otimes \underbrace{1 \otimes \cdots \otimes 1}_{(N-i-1) \text{ factors}}.$$

They generate the full permutation group [Lecture 10] and commute with the action of $\mathbb{C}G$. Each subspace of $V \otimes V \otimes \cdots \otimes V$ transforming by an irreducible representation of S_N is invariant under the action of $\mathbb{C}G$. They define particles with definite statistics. Bosons and fermions are obtained by the representations where $\tau_{i,i+1} \to \pm 1$ respectively. The corresponding vector spaces are the symmetrized and anti-symmetrized tensor products $V \otimes_{S,A} V \otimes_{S,A} \cdots \otimes_{S,A} V \equiv V^{\otimes s}, V^{\otimes A}$.

In quantum theory, there is the further assumption that *all observables commute with* $\tau_{i,i+1}$. Hence the above symmetrized and anti-symmetrized subspaces are invariant under the full observable algebra.

We can proceed as follows to generalize these considerations to any coproduct. The coproduct can be written as a series:

$$\Delta(\eta) = \sum_\alpha \eta_\alpha^{(1)} \otimes \eta_\alpha^{(2)} \equiv \eta_\alpha^{(1)} \otimes \eta_\alpha^{(2)}$$

$$\equiv \eta^{(1)} \otimes \eta^{(2)}.$$

Such a notation is called the "Sweedler notation". Then to every coproduct Δ, there is another coproduct Δ^{op} ('op' standing for 'opposite'):

$$\Delta^{\text{op}} = \eta_\alpha^{(2)} \otimes \eta_\alpha^{(1)} \equiv \eta^{(2)} \otimes \eta^{(1)}.$$

Suppose that Δ and Δ^{op} are equivalent in the following sense: There exists an *R-matrix* $R \in H \otimes H$ such that

i) R is invertible,

ii) $\Delta^{\mathrm{op}}(\eta)R = R\Delta(\eta)$,

and fulfills also certain further properties (Yang-Baxter relations) to which we will come later. Then the Hopf algebra is said to be "quasi-triangular".
For the simple case $\Delta(g) = g \otimes g$, $R = 1 \otimes 1$.

If i) is true, and ρ is the representation of H, $(\rho \otimes \rho)\Delta(\alpha)$ commutes with $\tau(\rho \otimes \rho)R$ as we now show. Therefore at least for two identical particles, we can use τR in place of τ to define statistics. More on this below. Let us first prove the result claimed. We have

$$(\rho \otimes \rho)\Delta(\eta)\tau\left[(\rho \otimes \rho)R\right] v \otimes w = (\rho \otimes \rho)\Delta(\eta)\tau \left[\rho(r_\alpha^{(1)})v\right] \otimes \left[\rho(r_\alpha^{(2)})w\right]$$

$$= \rho(\eta_\beta^{(1)} r_\alpha^{(2)})w \otimes \rho(\eta_\beta^{(2)} r_\alpha^{(1)})v \qquad (10)$$

while

$$\tau\left[(\rho \otimes \rho)R\right](\rho \otimes \rho)\Delta(\eta)v \otimes w = \tau(\rho \otimes \rho)\Delta^{\mathrm{op}}(\eta)\rho(r_\alpha^{(1)})v \otimes \rho(r_\alpha^{(2)})w$$

$$= \tau\rho(\eta_\beta^{(2)} r_\alpha^{(1)})v \otimes \rho(\eta_\beta^{(1)} r_\alpha^{(2)})w$$

$$= \rho(\eta_\beta^{(1)} r_\alpha^{(2)})w \otimes \rho(\eta_\beta^{(2)} r_\alpha^{(1)})v$$

which is Eq. (14). Here $R = r_\alpha^{(1)} \otimes r_\alpha^{(2)}$. The result is thus proved.

We can next decompose $V \otimes V$ into irreducible subspaces of τR. The observables are required to commute with τR. These irreducible subspaces describe particles of definite statistics.

We can generalize τR to N-particle sectors. Thus let

$$R_{i,i+1} = \underbrace{1 \otimes 1 \otimes \cdots \otimes 1}_{(i-1) \text{ factors}} \otimes R \otimes \underbrace{1 \otimes 1 \otimes \cdots \otimes 1}_{(N-i-1) \text{ factors}}.$$

Then

$$\mathcal{R}_{i,i+1} = \tau_{i,i+1}(\rho \otimes \rho \otimes \cdots \otimes \rho)R_{i,i+1}$$

generalizes $\tau_{i,i+1}$ of $\mathbb{C}G$ on $V \otimes V \otimes \cdots \otimes V (N\text{-factors})$. The group it generates replaces S_N. We will see below that it is the braid group $\mathcal{B}_N$. [See references on Hopf Algebras.]

As we really have different $\tau_{i,i+1}$ for different N, we sometimes call $\tau_{i,i+1}(\rho \otimes \rho \otimes \cdots \otimes \rho)R_{i,i+1}$ as $\rho \otimes \rho \otimes \cdots \otimes \rho[\tau_{i,i+1}R_{i,i+1}]$:

$$\tau_{i,i+1}(\rho \otimes \rho \otimes \cdots \otimes \rho)R_{i,i+1} \equiv \rho \otimes \rho \otimes \cdots \otimes \rho[\tau_{i,i+1}R_{i,i+1}].$$

Also, $\mathcal{R}_{i,i+1}$ depends on N, but for simplicity we do not show this dependence.

Remark.

The square of τR (or $\tau_{i,i+1}R_{i,i+1}$) is not necessarily identity. There could be representations where its eigenvalues are phases leading to 'anyons'.

Example

As a simple example, we consider the twisted coproduct Δ_θ,

$$\Delta_\theta(g) = F_\theta^{-1}(g \otimes g)F_\theta$$

for the Poincaré group.

Since τ commutes with $(\rho \otimes \rho)\Delta_0(g)$,

$$(\rho \otimes \rho)\Delta_0(g) = \tau(\rho \otimes \rho)[\Delta_0(g)],$$

one can see from the structure of $\Delta_\theta(g)$ that

$$\tau_\theta = (\rho \otimes \rho)F_\theta^{-1}\tau(\rho \otimes \rho)F_\theta$$

commutes with $(\rho \otimes \rho)\Delta_\theta(g)$. We can also write τ_θ as

$$\tau_\theta = \tau(\rho \otimes \rho)F_\theta^2 \tag{11}$$

since

$$(\rho \otimes \rho)F_\theta^{-1}\tau = \tau(\rho \otimes \rho)F_\theta. \tag{12}$$

From Eq. (11) we can deduce the R-matrix R_θ:

$$R_\theta = F_\theta^2.$$

We can also show this result by examining $\Delta_\theta^{\mathrm{op}}$. It is simple to show that

$$\Delta_\theta^{\mathrm{op}}(g) = F_\theta(g \otimes g)F_\theta^{-1}.$$

Clearly

$$R_\theta\Delta_\theta(g) = \Delta_\theta^{\mathrm{op}}(g)R_\theta$$

as we want.

A point to note is that

$$\tau_\theta^2 = \text{identity}$$

in view of Eq. (12). Hence the statistics group is still the permutation group S_N.

A Hopf algebra with S_N as statistics group is said to be "triangular".

LECTURE 43

i) The braid group and Yang-Baxter relations

The braid group $\mathcal{B}_N$ generalizes the statistics group S_N when the symmetry algebra is generalized to a "quasi-triangular" Hopf algebra. For the Hopf algebra to be quasi-triangular, it is necessary that R fulfills an additional relation called the Yang-Baxter relation. It can be derived by requiring that $\mathcal{R}_{i,i+1}$ generate $\mathcal{B}_N$. We now motivate the conditions on $\mathcal{R}_{i,i+1}$ which leads to $\mathcal{B}_N$ and then derive the Yang-Baxter relation.

The idea is simple. Consider

$$u \otimes v \otimes w \in V \otimes V \otimes V.$$

We can shuffle the left-hand side to its anti-cyclic form in two ways: First apply $\mathcal{R}_{12}$, then $\mathcal{R}_{23}$ and finally $\mathcal{R}_{12}$ again. On applying $\mathcal{R}_{12}$, we get

$$v_\alpha \otimes u_\alpha \otimes w,$$

repeated indices being summed. Next under $\mathcal{R}_{23}$, this becomes

$$v_\alpha \otimes w_\beta \otimes u_{\alpha,\beta}.$$

Finally on hitting with $\mathcal{R}_{12}$ once more, we find

$$w_{\beta,\alpha} \otimes v_{\alpha,\beta} \otimes u_{\alpha,\beta}.$$

The operator $\mathcal{R}_{23}\mathcal{R}_{12}\mathcal{R}_{23}$ also shuffles the left-hand side to the anti-cyclic form:

$$\mathcal{R}_{23} : u \otimes v \otimes w \rightarrow u \otimes w'_\alpha \otimes v'_\alpha$$

$$\mathcal{R}_{12}\mathcal{R}_{23} : u \otimes v \otimes w \rightarrow w'_{\alpha,\beta} \otimes u'_\beta \otimes v'_\alpha$$

$$\mathcal{R}_{23}\mathcal{R}_{12}\mathcal{R}_{23} : u \otimes v \otimes w \rightarrow w'_{\alpha,\beta} \otimes v'_{\alpha,\beta} \otimes u'_{\beta,\alpha}.$$

On requiring that the final result in either case is the same, we get

$$\mathcal{R}_{12}\mathcal{R}_{23}\mathcal{R}_{12} = \mathcal{R}_{23}\mathcal{R}_{12}\mathcal{R}_{23}. \tag{13}$$

In N-particle sectors, this generalizes to

$$\mathcal{R}_{i,i+1}\mathcal{R}_{i+1,i+2}\mathcal{R}_{i,i+1} = \mathcal{R}_{i+1,i+2}\mathcal{R}_{i,i+1}\mathcal{R}_{i+1,i+2}.$$

The group generated by $\mathcal{R}_{i,i+1}$ with these relations is known as the braid group $\mathcal{B}_N$. If the additional relation

$$\mathcal{R}^2_{i,i+1} = \text{identity}$$

is imposed, it becomes the permutation group S_N.

The Yang-Baxter Relations

These follow from the relation Eq. (13) on $\mathcal{R}_{i,i+1}$ when they are expressed in terms of $R_{i,i+1}$. We now show this.

Since $\tau_{i,i+1}$ generate S_N, they too fulfill

$$\tau_{i,i+1}\tau_{i+1,i+2}\tau_{i,i+1} = \tau_{i+1,i+2}\tau_{i,i+1}\tau_{i+1,i+2}$$

in addition to being idempotent.

Further

$$\tau_{ij} = \tau_{ji} = \tau_{ij}^{-1}, \quad i \neq j,$$

$$\tau_{ij}\rho(R_{i\beta}) = \rho(R_{j\beta})\tau_{ij}, \quad \text{if } \beta \notin \{i,j\},$$

$$\tau_{ij}\rho(R_{\alpha i}) = \rho(R_{\alpha j})\tau_{ij}, \quad \text{if } \alpha \notin \{i,j\}, \tag{14}$$

and

$$\tau_{ij}\rho(R_{ij}) = \rho(R_{ji})\tau_{ij}. \tag{15}$$

These identities are easily proved by acting on a generic vector in $V \otimes V \otimes \cdots \otimes V$. They express the fact that τ_{ij} changes either index of $\rho(R_{\alpha\beta})$ which is $i(j)$ to $j(i)$.

Now consider the braid relation

$$\tau_{i,i+1}\rho(R_{i,i+1})\tau_{i+1,i+2}\rho(R_{i+1,i+2})\tau_{i,i+1}\rho(R_{i,i+1})$$

$$= \tau_{i+1,i+2}\rho(R_{i+1,i+2})\tau_{i,i+1}\rho(R_{i,i+1})\tau_{i+1,i+2}\rho(R_{i+1,i+2}).$$

For the left-hand side (L.H.S.), on moving the τ's to the left extreme using Eq. (14), we get

$$\text{L.H.S.} = \tau_{i,i+1}\tau_{i+1,i+2}\tau_{i,i+1}\rho(R_{i+1,i+2}R_{i,i+2}R_{i,i+1}).$$

Similarly for the right-hand side (R.H.S.), we find

$$\text{R.H.S.} = \tau_{i+1,i+2}\tau_{i,i+1}\tau_{i+1,i+2}\rho\big(R'_{i,i+1}R_{i,i+2}R_{i+1,i+2}\big).$$

Therefore

$$\rho\big(R_{i+1,i+2}R_{i,i+1}R_{i,i+1}\big) = \rho\big(R_{i,i+1}R_{i,i+2}R_{i+1,i+2}\big).$$

This is correct in any representation ρ. So it is natural to require that it holds for R_{ij} itself, regarding it as an element of $H \otimes H \otimes \cdots \otimes H$.

The relations

$$R_{i+1,i+2}R_{i,i+2}R_{i,i+1} = R_{i,i+1}R_{i,i+2}R_{i+1,i+2}$$

are known as the (quantum) Yang-Baxter relations. For $i = 1$, they read

$$R_{23}R_{13}R_{12} = R_{12}R_{13}R_{23}.$$

Remark:

There is another way to derive this relation. Thus Δ^{op} is $R\Delta R^{-1}$ and both Δ and Δ^{op} satisfy the identity coming from coassociativity. That imposes a condition on R. It is fulfilled when R satisfies the Yang-Baxter relation.

LECTURE 44

i) The Kuperberg diagrams

Hopf algebras involve the structures m, Δ, ϵ and S which satisfy a number of axioms. The diagramatic notations of Kuperberg make them more transparent. We describe this notation in this section following Kuperberg (G. Kuperberg, Duke Math. J. 84 (1996)83-129)[arXiv: q-alg/9712047] and L. H. Kauffman and D. E. Radford [arXiv: math/9911081v1(math. QA)].

There is one more mapping η which it is convenient to add to the above list. The Hopf algebras H we consider are unital, so they contain $\mathcal{C}1$, an isomorph of $\mathcal{C}$. The map η sends $\zeta \in \mathcal{C}$ to $\zeta 1$ in H. This map is called the *unit*. For a physicist, the identification established by η is obvious.

Together with H, we now consider also H^*, the vector space dual of H.

Remark The dual H^* is also canonically a Hopf algebra by duality, with $H^{**} = H$ (certainly if H is finite dimensional, or else with additional technical assumptions). Together, $H \otimes H^*$ is also a Hopf algebra, the Drinfel'd double of H. [For details see the references on Hopf Algebras.]

Each of the maps m, Δ, ϵ, S, η is now to be regarded as a tensor. So they are elements of $H^{\otimes k} \otimes H^{* \otimes l}$ where

$$V^{\otimes k} := \underbrace{V \otimes V \otimes \cdots \otimes V}_{k \text{ factors}}.$$

The number of output arrows on a letter in the diagrams below is the value of k, the number of input arrows shows the value of l. We adopt the convention that $H^{\otimes 0} = H^{* \otimes 0} = \mathcal{C}$. When $k(l)$ is zero, the corresponding input(output) arrow may not be shown.

Thus since

$$m: \quad H \otimes H \to H,$$

$$m \in H \otimes H^{*2},$$

and m appears in the diagrams with two input and one output arrow:

Since

$$\Delta: \quad H \to H \otimes H,$$

it appears with one input and two output arrows:

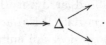

The notation for ϵ is

$$\longrightarrow \epsilon \, .$$

The output for ϵ has $k = 0$ or zero output line, and hence it is not shown. Next comes S:

$$\longrightarrow S \longrightarrow \, .$$

As for the unit η, which maps $\mathcal{C}$ to $\mathcal{C}1 \in H$, it has $l = 0$ and hence is represented by

$$\eta \longrightarrow \, .$$

Elements a of H are represented by just one output arrow just like η:

$$a \longrightarrow \, .$$

This means that we identify η as $1 \in H$ whereas before, $\eta: \mathcal{C} \to H$. No error can come from this dual use of η, as $a \in H$ can also be thought of as a linear map $\hat{a}: \mathcal{C} \to H$.

Tensor products are denoted by juxtaposition. Thus

$$a \longrightarrow$$

$$b \longrightarrow$$

denotes $a \otimes b$.

We can now complete the decoration of incoming and outgoing lines of m and Δ:

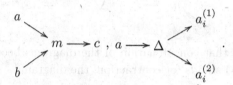

The first diagram stands for

$$m : a \otimes b \to c$$

or $ab = c$. The second stands for

$$\Delta : a \to a_i^{(1)} \otimes a_i^{(2)}$$

(with i summed over) or $\Delta(a) = a^{(1)} \otimes a^{(2)}$ in Sweedler notation.

We also need rules for concatenation of diagrams. Consider

$$a \longrightarrow b .$$

Since $a \in H$ and $b \in H^*$, the natural interpretation is that this is the contraction $b(a)$.

We can also interpret this using a basis $e_i \in H$ (assuming that H has a nice basis, say that H is finite-dimensional) and the dual basis $e^i \in H^*$:

$$e^i(e_j) = \delta_j^i.$$

Then

$$a = a^i e_i, \quad b = b_j e^i, \quad a^i, b_j \in \mathcal{C}$$

and

$$b(a) = a^i b_i.$$

We can make this explicit by decorating the diagram with indices:

$$a \underset{i}{\longrightarrow} b . \qquad\qquad (*)$$

If $m_{i,j}^k$ and $\Delta_i^{j,k}$ are the structure constants of the algebra and coalgebra in the above basis,

$$m(e_i \otimes e_j) = m_{i,j}^k e_k, \quad \Delta(e_i) = \Delta_i^{j,k} e_j \otimes e_k,$$

the $m-$ and $\Delta-$ diagrams can also be decorated with indices as follows:

$$m_{i,j}^k \quad = \quad \underset{j}{\overset{i}{\searrow}}\ m \xrightarrow{\ k\ }, \quad \Delta_i^{j,k} \quad = \quad \xrightarrow{\ i\ } \Delta \overset{\overset{j}{\nearrow}}{\underset{\underset{k}{\searrow}}{}}.$$

The advantage is that concatenation of the diagrams becomes easy. Just as the diagram ($*$) stands for contraction, the diagram

$$L$$

stands for tr L (as repeated indices are summed) while

$$\underset{j}{\overset{i}{\searrow}}\ m \xrightarrow{\ k\ } \Delta \overset{\overset{l}{\nearrow}}{\underset{\underset{m}{\searrow}}{}}$$

stands for $m_{i,j}^k \Delta_k^{l,m}$.

Note that incoming indices at each vertex (here i, j) are to be read *counter-clockwise* and outgoing indices (here l, m) are to be read clockwise. This is important since for example $m_{i,j}^k \neq m_{j,i}^k$ if H is not a commutative algebra.

Since

$$a \longrightarrow$$

stands for $a \in H$, Kuperberg interprets $\rightarrow$ to stand for the identity linear transformation. He interprets

$$\bigcap$$

to stand for

$$\bigcap \\ 1 \ = \ \mathrm{Tr}\ 1$$

or the dimension of H.

As an example, let us write condition

$$\Delta(ab) = \Delta(a)\Delta(b).$$

It is

$$(**)$$

We can decorate the lines with indices,

$$(***)$$

meaning

$$m_{i,j}^p \Delta_p^{k,l} = \Delta_i^{p,q} \Delta_j^{r,s} m_{p,r}^k m_{q,s}^l.$$

It can be checked that this is the relation

$$\Delta(e_i e_j) = \Delta(e_i)\Delta(e_j).$$

We can also decorate it with generic elements of H using Sweedler's notation that gives us a basis-free interpretation.

We can now state the axioms of a Hopf algebra H using diagrams.

A Hopf algebra H consists of the tensors m, Δ, η, ϵ and S called the product or multiplication, the coproduct or comultiplication, the unit, the counit and the anti-pode respectively. They are represented by the diagrams

$$
\searrow\!\!\!\!\nearrow m \longrightarrow \;,\quad \longrightarrow \Delta \!\!\!\begin{smallmatrix}\nearrow\\\searrow\end{smallmatrix}\;,\quad \eta \longrightarrow\;,\quad \longrightarrow \epsilon\;,\quad \longrightarrow S \longrightarrow \;.
$$

Multiplication is associative and unital ("unital" here implying that H has unit η):

$$
\begin{array}{c}\searrow\!\!\!\nearrow m \\ \searrow\!\!\!\nearrow\end{array} m \longrightarrow \;=\; \begin{array}{c}\searrow m \\ \nearrow\;\;\searrow\!\!\!\nearrow\end{array} m \longrightarrow \;,
$$

$$
\begin{array}{c}\eta\searrow \\ \;\;\nearrow\end{array} m \longrightarrow \;=\; \begin{array}{c}\searrow \\ \nearrow\;\eta\end{array} m \longrightarrow \;.
$$

Recall that comultiplication is a homomorphism from H to $H \otimes H$. See (**) and (* * *). It is coassociative and counital.

$$
\longrightarrow \Delta \begin{array}{c}\nearrow \\ \searrow\;\nearrow \\ \;\;\Delta\searrow\end{array} \;=\; \begin{array}{c}\nearrow\Delta\nearrow \\ \longrightarrow\Delta\searrow\;\searrow\end{array} \;,
$$

$$
\longrightarrow \Delta \begin{array}{c}\nearrow\epsilon \\ \searrow\end{array} \;=\; \longrightarrow \Delta \begin{array}{c}\nearrow \\ \searrow\epsilon\end{array} \;.
$$

The tensors are related by the bialgebra axiom (∗∗) and the axiom of the antipode:

$$= \longrightarrow \epsilon \cdot \eta \longrightarrow \ \cdot$$

In this last diagram, the output of ϵ is in $\mathbb{C}$ which then is mapped back to H by η.

We can interpret these diagrams in equations by generic inputs and equating outputs. For example

means

$$m(\eta \otimes a) = a$$

and

means

$$S(a_i^{(1)})a_i^{(2)} = \epsilon(a)1$$

while $\Delta(a) = a_i^{(1)} \otimes a_i^{(2)}$.

Remarks.

The above diagrams do not incorporate the ∗ operation and quasi-triangularity. They are additional structures on H.

This concludes our discussion on Hopf algebras. Literature may be consulted for further mathematical discussions and physical applications.

PROBLEMS

General Notions

1) A transposition $(i\ j)$ in S_n is an element of the form

$$\begin{pmatrix} 1\ 2\ \cdots\ i-1\ i\ i+1\ \cdots\ j-i\ j\ j+1\ \cdots \\ 1\ 2\ \cdots\ i-1\ j\ i+1\ \cdots\ j-1\ i\ j+1\ \cdots \end{pmatrix}.$$

It permutes only i and j. Any element of S_n is a product of transpositions. The cyclic permutation $(1\ 2\ 3\ \cdots\ n)$ is the element

$$\begin{pmatrix} 1\ 2\ 3\ \cdots\ n-1\ n \\ 2\ 3\ 4\ \cdots\ n\ \ \ 1 \end{pmatrix}.$$

(a) Show that $(1\ 2\ 3\ \cdots\ n)(i\ i+1)(1\ 2\ 3\ \cdots\ n)^{-1} = (i+1\ i+2)$.
(b) Show that $(k\ k+1)(i\ k)(k\ k+1)^{-1} = (i\ k+1)$.

2) A subset G_0 in a group G is a subgroup of G if G_0 itself is a group under the group composition law defined on G. Show that if H and K are subgroups of a group G, then $H \cap K$ is a subgroup of G.

3) The group $\mathcal{E}_2$ of Euclidean motions in a plane (i.e. $\mathbb{R}^2$) is defined as follows: An element of $\mathcal{E}_2$ is the pair $(a,\ R(\phi))$ where $a \in \mathbb{R}^2$ and

$$R(\phi) = \begin{bmatrix} \cos\phi & -\sin\phi \\ \sin\phi & \cos\phi \end{bmatrix}$$

is a rotation. $\mathcal{E}_2$ acts as a transformation group on $\mathbb{R}^2$ by the action $(a,\ R(\phi))x = R(\phi)x + a$ for all $x \in \mathbb{R}^2$. (a) Infer the group composition law of $\mathcal{E}_2$. (b) Show that the subsets $\{(a,\ 1)\}$ and $\{(0,\ R(\phi))\}$ form subgroups of $\mathcal{E}_2$.

4) (a) Let

$$t(a) = \begin{bmatrix} 1 & a_1 + ia_2 \\ 0 & \cdot \; 1 \end{bmatrix}$$

for each $a = (a_1, a_2) \in \mathbb{R}^2$. Show that $t(a)$'s generate a group under matrix multiplication isomorphic to the subgroup of translations in $\mathcal{E}_2$.
(b) Let

$$d(\phi) = \begin{bmatrix} e^{i\phi/2} & 0 \\ 0 & e^{-i\phi/2} \end{bmatrix}, \quad 0 \le \phi \le 4\pi.$$

Show that the $d(\phi)$'s generate a group under matrix multiplication which homomorphically covers the subgroup of rotations in $\mathcal{E}_2$ twice for a suitable choice of the homomorphism. ["Twice" here means that there are precisely two distinct $d(\phi)$'s which get mapped into the same rotation.]
(c) Consider the set of all matrices

$$t(a)\, d(\phi) = \begin{bmatrix} e^{i\phi/2} & (a_1 + ia_2)e^{-i\phi/2} \\ 0 & e^{-i\phi/2} \end{bmatrix}.$$

Show that they form a group under multiplication which homomorphically covers $\mathcal{E}_2$ twice.
Hint: $t(a)\, d(\phi)\, t(a')\, d(\phi') = \{t(a)\, d(\phi)\, t(a')\, d(\phi)^{-1}\}\, d(\phi)\, d(\phi')$
and $d(\phi)\, t(a')\, d(\phi)^{-1} = t[R(\phi)a']$.
(d) Show that translations form an invariant subgroup of $\mathcal{E}_2$.

5) Let $U(1) \times SU(n) = \{(g,\, h)\}_{g \in U(1), h \in SU(n)}$. Show that $(g,\, h) \to gh$ is a homomorphism from $U(1) \times SU(n)$ onto $U(n)$. [The product gh has an obvious meaning.] Show that the kernel of the homomorphism is a discrete subgroup of order n. [Thus $U(1) \times SU(n)$ covers $U(n)$ n times.]

6) Show that : (a) If H is a subgroup of a group G and N is a *normal* (i.e. invariant) subgroup of G, then $N \cap H$ is a *normal* subgroup of H.
(b) If H is a subgroup of a group G and N is a *normal* subgroup of G, then $NH = \{nh | n \in N,\ h \in H\}$ is a subgroup of G.
(c) If H and N are normal subgroups of a group G, then NH and $N \cap H$ are normal subgroups of G.

7) What are the elements conjugate to

$$\begin{pmatrix} 1 & 2 & \cdots & n \\ s_1 & s_2 & \cdots & s_n \end{pmatrix}$$

in S_n?

8) Exhibit the invariant subgroup of $SU(2)$ which is neither the whole group nor consists of identity alone.

9) Let $f : G \to G'$ be a homomorphism from a group G onto a group G'. Let Ker f denote the kernel of the homomorphism. Show that Ker f is an invariant subgroup of G.

10) The center C of a group G is the set $\{g\}$ of all elements in G which commute with every element in G.
(a) Show that C is an invariant subgroup of G.
(b) The group $SU(n)$ is the set of all $n \times n$ unitary matrices with determinant 1. What is the center of $SU(n)$? You may assume that if a matrix M commutes with all the elements of $SU(n)$, then M is a multiple of the unit matrix.

11) Let f be a homomorphism from a group G onto a group G' with kernel Ker f.
(a) Show that the factor group $G/Ker\ f$ is isomorphic to G'.
(b) Show that there is an invariant subgroup K of $SU(2)$ such that $SU(2)/K$ is isomorphic to $SO(3)$.

12) We know that in $\mathcal{E}_2$, the translations $T_2 = \{(a,\ 1)\}$ in two dimensions form an invariant subgroup. Find the group of matrices isomorphic to the factor group $\mathcal{E}_2/T_2$.

13) Let the group G be simple. Let f be a homomorphism from G onto a group G'.
(a) Show that either f is an isomorphism or G' consists of identity alone.
(b) Hence show that all representations of a simple group, except the trivial one where all group elements are represented by the unit matrix, are faithful. *Note*: $SO(3)$ and the proper orthochronous Lorentz group $\mathcal{L}_+^\uparrow$ are simple. But $SU(2)$ and $SL(2,\ \mathbf{C})$ are not simple. Can you see the reason for the latter?

14) Suppose that the only one-dimensional representation of a group is the trivial one. Let $\Gamma = \{D(g)\}$ be a finite dimensional representation of G.

Show that det $D(g) = 1$ for $g \in G$.

15) Let G be a group, with elements $\{g_\alpha\}$. The element $q = g_1 g_2 g_1^{-1} g_2^{-1}$ is called the commutator of g_1 and g_2. Let Q be the set of all elements of the form $q_1 q_2 \cdots q_m$ where q_i is any commutator and m is any positive integer. Show that Q is an invariant subgroup of G. [Q is the *first commutator* or *first derived* group of G.]

16) Let G be a group and Q its commutator subgroup.
(a) Show that G/Q is Abelian. [Hint: For A, $B \in G/Q$, show that $ABA^{-1}B^{-1} \cap Q \neq \emptyset$. Show that $ABA^{-1}B^{-1}$ is a coset. Hence infer that $ABA^{-1}B^{-1} = Q$ and $AB = BA$.]
(b) Show that if N is a normal subgroup of G, and G/N is Abelian, then $N \supseteq Q$. [Hint: If $N \not\supseteq Q$, then N cannot contain all commutators. If $aba^{-1}b^{-1} \notin N$, show that $(aN)(bN)(aN)^{-1}(bN)^{-1} \neq N$ to infer the result.]

17) Suppose we are given two groups G and H and a homomorphism τ of H into the group Aut G of automorphisms of G. [Aut G is clearly a group under the natural composition law.] Thus for $h \in H$, $\tau(h)$ is an isomorphism of G onto G so that $\tau(h) : G \to G$ is $1 - 1$ onto with $\tau(h)\,(g_1 g_2) = [\tau(h)g_1][\tau(h)g_2]$. The *semi-direct product* of G with H are the pairs (g, h) [$g \in G$, $h \in H$] with the multiplication law $(g_1, h_1)(g_2, h_2) = (g_1 \tau(h_1)g_2, h_1 h_2)$. Show that the semi-direct product of G with H is a group which has $\{(g, e)\}_{g \in G}$ as an invariant subgroup. [Note $\{(g, e)\}_{g \in G}$ is isomorphic to G.] Remark: Often τ is omitted in the semi-direct product notation.

18) Let $\mathcal{P} = \{(a, \Lambda)\}$ be the Poincaré group with $[\tau(\Lambda)a]^\mu = \Lambda^\mu{}_\nu a^\nu$. Show that $\mathcal{P}$ is the semi-direct product of translations in $\mathbb{R}^4$ with Lorentz transformations.

19) For $SU(2)$ [or $SO(3)$], an irreducible representation of spin j takes the form $SU(2) \ni g \to D(g) = e^{i \sum J_i \theta_i}$ for appropriate coordination of g, where θ_i are real and J_i are spin j angular momentum matrices. It is also known that J_1 and J_3 are real and J_2 is pure imaginary for any j. [More precisely they can be chosen to be so by an appropriate choice of basis.] Show that for every j, the representations $\{D(g)\}$ and $\{D^*(g)\}$ are equivalent. Hint:

Think of the matrix $C = e^{i\pi J_2}$.

FINITE GROUPS

20) For S_3, show that any two distinct transpositions form a system of generators.

21) Let G be a finite group of order n. Let $\Gamma_i = \{T_i(g)\}$ be the left regular representation of G obtained by converting G into a vector space V_1 and considering left action on it. Let $\Gamma_2 = \{T_2(g)\}$ be the representation of G (also called left regular) obtained by considering the vector space V_2 of all functions from G into complex numbers and considering the left action on these functions described in the text. Exhibit a basis for V_1 and V_2 so that $T_1(g)$ and $T_2(g)$ have identical matrices. [Hence Γ_1 and Γ_2 are equivalent. Hence too the identity of their names.] Hint: Show that the functions $f_\sigma(\sigma = 1, 2, \cdots, n)$ defined by $f_\sigma(g_p) = \delta_{\sigma p}$ form a basis for V_2.

22) Consider the left regular representation on the vector space V of functions from a group G (finite or not) into complex numbers. Show that *every* finite dimensional unitary irreducible representation Γ of G is equivalent to the left regular representation when it is restricted to act on a suitable subspace V_Γ of V. You may assume that if $\Gamma = \{D(g)\}$, then the *functions* D_{ij}^* in V are linearly independent i.e $\sum d_{ij}D_{ij}^* = 0 \Rightarrow d_{ij} = 0$ [Thus every finite dimensional unitary irreducible representation occurs in the reduction of the left regular representation of a group.] Hint: The linear manifold spanned by D_{ij}^* for linear combinations on either index is a subspace of V. Consider the action of the left regular representation on these functions.

23) Let $\Gamma_\alpha = \{D^\alpha(g)\}$ denote all the inequivalent unitary irreducible representations (UIRR's) of a finite group G of order n. Let the matrices D^α be $\sigma_\alpha \times \sigma_\alpha$ i.e dim Γ_α = dim of vector space on which Γ_α operates = σ_α. It is known, and you may assume, that $\sum_\alpha \sigma_\alpha^2 = n$.

Let V be the vector space of functions from G into complex numbers. (a) Show that the functions $D_{ij}^* \in V$ are linearly independent and hence form a basis for V. In the expansion $V \ni f = \sum_{\alpha,i,j} \xi_{ij}^\alpha D_{ij}^{\alpha*}$, find a formula for ξ_{ij}^α in terms of f and D_{ij}^α. [This is the Peter-Weyl theorem for finite groups, the expansion being the "Fourier" series for functions from a finite group into complex numbers.]

(b) Show that

$$V = \oplus_\alpha \oplus_{j=1}^{\sigma_\alpha} V_\alpha^i$$

where the left regular representation on $V_\alpha^i (i = 1, 2, \cdots, \sigma_\alpha)$ is equivalent to Γ_α by exhibiting a basis for each V_α^i. [Thus, the left regular representation Γ on V is $\oplus_\alpha \sigma_\alpha \Gamma_\alpha$ where by $\sigma_\alpha \Gamma_\alpha$ we mean

$$\underbrace{\Gamma_\alpha \oplus \Gamma_\alpha \oplus \cdots \oplus \Gamma_\alpha}_{\sigma_\alpha \text{ times}}$$

and equivalent representations are identified. Note that in the reduction of Γ into UIRR's, each Γ_α of dimension σ_α occurs σ_α times.]

24) For S_4, find an invariant subgroup G_0 such that S_4/G_0 is S_3.

25) Let $\bar{V}$ be the vector space a basis for which is the group elements of a finite group G of order n. Let V be the vector space of left regular representation on complex functions on G. We have seen that $f_i(i = 1, \cdots, n) \in V$ form a basis for V where $f_i(g_j) = \delta_{ij}$ and that the map $g_i \to f_i$ induces an isomorphism between the left regular representations on $\bar{V}$ and V. Show that if $g_i \to f_i(i = 1, \cdots, n)$, then $g_j g_i \to f_{ji}$ where $f_{ji}(g_\rho) = \sum_\alpha f_j(g_\alpha) f_i(g_\alpha^{-1} g_\rho)$. [$f_{ji} = f_j * f_i$ is called the *convolution* of functions f_j and f_i. Thus the group algebra is isomorphic to the convolution algebra of functions from G to complex numbers.]

26) Let G be a finite group of order n, $\Gamma_\alpha = \{D^\alpha(g)\}_{g \in G}$ its inequivalent unitary IRR's. If $f_1 = \sum_{\alpha,i,j} \xi_{ij}^\alpha D_{ij}^{\alpha*}$, $f_2 = \sum_{\alpha,i,j} \eta_{ij}^\alpha D_{ij}^{\alpha*}$ are the "Fourier" series for the two functions f_1, f_2 from G into complex numbers, and if $f_1 * f_2 = \sum_{\alpha,i,j} \zeta_{ij}^\alpha D_{ij}^{\alpha*}$ [$f_1 * f_2$ being the convolution of f_1 and f_2], find a formula for ζ_{ij}^α in terms of ξ_{ij}^α and η_{ij}^α.

27) Let G be a finite group with inequivalent irreducible representations (IRR's) $\Gamma_\alpha = \{D^\alpha(g)\}$ and associated characters $\{\chi_\alpha(g)\}$. Show that $\chi_\alpha(g) \chi_\beta(g) = \sum_\gamma N_{\alpha\beta}^\gamma \chi_\gamma(g)$ where $N_{\alpha\beta}^\gamma$ are non-negative integers. [$N_{\alpha\beta}^\gamma$ are sometimes called "fusion coefficients".]

28) Find all the matrices of the two-dimensional IRR of S_3 taking

$$\Gamma(\{2,\ 1\})_1 = \begin{array}{|c|c|} \hline 1 & 2 \\ \hline 3 \\ \cline{1-1} \end{array}, \quad \Gamma(\{2,\ 1\})_2 = \begin{array}{|c|c|} \hline 1 & 3 \\ \hline 2 \\ \cline{1-1} \end{array}$$

in the notation of the text. Write the identity $e \in S_3$ as an appropriate sum of Young symmetrizers.

29) Find the dimensions σ_α of IRR's of S_4 using
(a) Young tableaus methods,
(b) the relation $4! = \sum_{\alpha=1}^{m} \sigma_\alpha^2$, m = number of classes of S_4.

LIE GROUPS

30) Let a, b, c, $\cdots$ be the coordinates of an n-dimensional Lie group G with $0 \in \mathbb{R}^n$, labeling the identity in G. Thus for each $a = (a^1, a^2, \cdots, a^n) \in \mathbb{R}^n$, $g(a) \in G$ and $g(0) = e$. Let $g(a)\,g(b) = g[\phi(a, b)]$ denote the group composition law where $\phi(a, b) = (\phi^1(a, b), \phi^2(a, b), \cdots, \phi^n(a, b))$. Let

$$\phi^\rho_{.\tau}(a, b) = \frac{\partial}{\partial b^\tau}\phi^\rho(a, b) \quad \text{and} \quad \phi^\rho_{\tau.}(a, b) = \frac{\partial}{\partial a^\tau}\phi^\rho(a, b).$$

Show from the associative law $g(a)[g(b)\,g(c)] = [g(a)\,g(b)]g(c)$ that $\det(\phi^\rho_{.\tau}(a, b))\,\det(\phi^\rho_{.\tau}(b, o)) = \det(\phi^\rho_{.\tau}[\phi(a, b), o])$. Hence show that if

$$d\mu_L(g) = \rho[g(a)]da^1 da^2 \cdots da^n$$

where

$$\rho[g(a)] = [\det(\phi^\rho_{.\tau}(a, o))]^{-1},$$

then $d\mu_L(g_0 g) = d\mu_L(g)$ for every fixed $g_0 \in G$. [This is the left invariance of the measure $d\mu_L(g)$.]

31) The group $G = \{(a_1, a_2)\}$ $(0 < a_1 < \infty, -\infty < a_2 < \infty)$ of translations and dilatations in 1 dimension is defined by the action $(a_1, a_2)x = a_1 x + a_2$ for all $x \in \mathbb{R}^1$. Show that for this group, the left and right invariant measures are not constant multiples of each other.

32) Show that every finite dimensional representation of a compact Lie group is equivalent to a unitary representation. [It follows that such representations are completely reducible.]

33) Let $\Gamma^\alpha = \{D^{(\alpha)}(g)\}$, $\alpha = 0, 1, \cdots$ denote all the inequivalent unitary irreducible representations of a compact Lie group G with invariant measure $d\mu(g)$ normalized as $\int_G d\mu(g) = V$. The dimension of $\Gamma^{(\alpha)}$ is σ_α. Show that

$$\int_G d\mu(g)\, D^\alpha_{ij}(g)\, D^\alpha_{kl}(g)^\dagger = \frac{V}{\sigma_\alpha}\,\delta_{\alpha\beta}\,\delta_{il}\,\delta_{jk}.$$

Hence derive the corresponding character orthogonality and normalization formulae.

34) Show that the left regular representation of a Lie group G is unitary in the Hilbert space H of functions from G to complex numbers $\mathbf{C}$ with the scalar product

$$(b_1,\ b_2) = \int_G d\mu_L(g)\ b_1^*(g)\ b_2(g).$$

Here $d\mu_L(g)$ is the left invariant measure.

35) Let $\Gamma^\alpha = \{D^{(\alpha)}(g)\}$, $\alpha = 1,\ 2,\ \cdots$ denote all the inequivalent UIRR's of a compact Lie group G with dimension dim $\Gamma^\alpha = \sigma_\alpha$. Let H be the Hilbert space carrying the left regular representation of G. The scalar product in H is given by

$$(b_1,\ b_2) = \int_G d\mu(g)\ b_1^*(g)\ b_2(g).$$

$d\mu$ being the invariant measure for G. Assume Perter-Weyl theorem according to which any $f \in H$ has the expansion

$$f(g) = \sum_{\alpha,i,j} \eta_{ij}^\alpha\ D_{ij}^{\alpha*}(g)$$

[the equality being understood in the sense of convergence in norm]. Show that in the reduction of the left regular representation on H, the UIRR Γ^α of dimension σ_α occurs with multiplicity σ_α.

36) An irreducible tensor operator T with angular momentum J for $SO(3)$ is a $(2J+1)$-tuple set of linear operators $(T^{-J},\ T^{-J+1},\ \cdots,\ T^J)$ with the transformation law $U(R)T^q U(R)^{-1} = D_{q'q}^J(R)T^{q'}$ where $\{D^J(R)\}$ is the $(2J+1)$-dimensional IRR of $SO(3)$ and $U(R)$ is the unitary operator which implements the rotations. Thus $U(R)|Jm\rangle = |Jm'\rangle D_{m'm}^J(R)$ in an obvious notation. Show that for $J \neq 0$,

$$\sum_{m=-J}^{+J} \langle Jm|T^q|Jm\rangle = 0 \quad \text{for } q = -J,\ -J+1,\ \cdots,\ +J.$$

37) Let L be a Lie algebra with basis $L_1,\ L_2,\ \cdots,\ L_n$. Define a set of linear operators $T(\alpha)$ on L by $T(\alpha)L_i = [L_\alpha,\ L_i]$. a) Show that $L_\alpha \to T(\alpha)$ (and of course $\sum_\alpha \xi_\alpha L_\alpha \to \sum_\alpha \xi_\alpha T(\alpha)$) defines a representation of L. b)

Hence show that the structure constants, understood suitably as a certain family of matrices, yield a representation of L. *Note*: The above is the *adjoint* representation of L.

38) Find a basis for the Lie algebras of the following groups and their commutation relations: a) $SO(n)$, b) $SU(1,1)$, c) $SO(2,1)$. Show that the Lie algebras of $SU(1,1)$ and $SO(2,1)$ are isomorphic.

39) Let $\Gamma = \{ad\ell\}$ denote the adjoint representation of the Lie algebra $L = \{\ell\}$. Find the conditions under which Γ is a faithful representation of L. The set of all $\ell \in L$ with $ad\ell = 0$ is called the kernel of the map from L to Γ. When the representation Γ is not faithful, how will you characterize the kernel of this map?

40) Show that the Lie algebras of $SO(4)$ and $SU(2) \otimes SU(2)$ are isomorphic. Find two independent Casimir operators for $SO(4)$ or $SU(2) \otimes SU(2)$. Characterize (i.e. label) the UIRR's of $SU(2) \otimes SU(2)$ in terms of the eigenvalues of these Casimir operators. [Be guided in your considerations by your knowledge of $SU(2)$ in doing the problem.]

41) Let $[T(g_0)f](x) = f(g_0^{-1}x)$ define a representation $SO(3)$ via $SO(3) \ni g_0 \to T(g_0)$. Here $x \in \mathbb{R}^3$ and f is a function from $\mathbb{R}^3$ to $\mathbb{C}$ while $g_0^{-1}x$ denotes the matrix g_0^{-1} acting on x. For small rotations, we may write $g_0 \simeq 1 + i\,\varepsilon_\alpha\theta_\alpha + \cdots$, where $i\theta_\alpha$ spans the Lie algebra of $SO(3)$ in the 3-dimensional representation: $(\theta_\alpha)_{ij} = -i\varepsilon_{\alpha ij}$. Similarly, we may write $T(g_0 \simeq 1 + i\varepsilon_\alpha\theta_\alpha + \cdots) = 1 + i\varepsilon_\alpha J_\alpha + \cdots$ where J_α is the representative of θ_α on the space of functions on $\mathbb{R}^3$. Show that $(J_\alpha f)(x) = -i\varepsilon_{\alpha ij}x_i\frac{\partial}{\partial x_j}f(x)$, i.e. that $\vec{J} = (\vec{r} \times \vec{p})$ in the conventional notation.

42) In the reduction of the representation of the Lie algebra of $SO(3)$ in 41) above, the carrier space for IRR's is spanned by functions $\{H_{lm}\}$, $m = -l, -l+1, \cdots, +l$, different l's giving different IRR's. The spectrum of l is $l = 0, 1, 2, \cdots$. What are the differential equations defining $\{H_{lm}\}$? State (need not demonstrate the statement) what H_{lm}'s are in terms of known special functions.

43) (a) Find a basis and the Lie brackets of its elements for the following groups: i) $SL(2, \mathbb{C})$, ii) the component $\mathcal{L}_+^\uparrow$ connected to identity of the

Lorentz group.

(b) Hence show that these Lie algebras are isomorphic.

44) (a) For $SU(2)$, show the decomposition of a tensor of rank three in terms of irreducible tensors. What are the "angular momentum" values of these irreducible tensors?

(b) For $SO(3)$, show the decomposition of a tensor of rank three in terms of irreducible tensors. What are the "angular momentum" values of these irreducible tensors?

45) Show that $SU(3)$ has IRR's of dimensions 6 and 8.

46) For $SU(2) \otimes SU(2)$, show that all IRR's are unfaithful. Show also that there are infinitely many reducible representations which are faithful.

47) Prove that there are no homomorphisms $SO(3) \times SU(2) \to SO(4)$ and $SO(4) \to SO(3) \times SU(2)$.

THE POINCARÉ GROUP

48) We say that a group G acts *transitively* on a space X and X is a *homogeneous space* for G if an action of G on X [denoted by $X \ni x \xrightarrow{\ g \in G\ } gx \in X$] is given such that $g_1(g_2 x) = (g_1 g_2)x$, and for each $x,\ y \in X$, there exists a $g \in G$ such that $y = gx$.

(a) Show that if X is a homogeneous space for G, then each $g \in G$ induces a map from X *onto* X, i.e. $gX = X$ symbolically.

(b) Consider the action of a group G on itself given by group multiplication. Thus each $g_0 \in G$ acts as the transformation $g \to g_0 g$ on G. Show that this is a transitive action.

(c) Give an example of a homogeneous space for $SO(3)$ other than $SO(3)$ itself.

49) Let G be a group of transformations on a space X. The *orbit* or *surface of transitivity* of each $x \in X$ is the set $Gx = \{gx | x \in G\}$. The orbit of x is clearly a homogeneous space containing x. a) Show that $X = \cup_\alpha X_\alpha$ where each X_α is a homogeneous space and $X_\alpha \cap X_\beta = \emptyset$. b) Consider the standard action of $SO(3)$ as linear transformations on $\mathbb{R}^3$. Write $\mathbb{R}^3$ as the union of orbits for such an action. *Note*: The answer for b) will show that

a homogeneous space in general is not a vector space.

50) Let G be a group and X a homogeneous space for G. Let G_x be the subset of all $g \in G$ such that $gx = x$, i.e. $G_x = \{g | g \in G,\ gx = x\}$. G_x is a subgroup of G, the *little, stationary* or *isotropy* group at x. Prove that:
(a) The little groups G_x at different $x \in X$ are conjugate to each other (and hence are isomorphic).
(b) For any one fixed $x \in X$ and the corresponding little group G_x, the left cosets $\{gG_x\}_{g \in G}$ are mapped in a $1 - 1$ way onto X by the map $gG_x \to gx \in X$.

51) Let $G = SO(3)$ and $X = S^2$, the surface of a unit radius in $\mathbb{R}^3$ centered around the origin. Then for the usual action of $SO(3)$ on $\mathbb{R}^3$, S^2 is a homogeneous space. a) What is the little group at any $x \in X$ isomorphic to? b) Let G_0 be the little group at $(0,\ 0,\ 1) \in S^2$. Using the Euler parametrization of $SO(3)$ and polar coordination of S^2, exhibit the $1 - 1$ map from $\{gG_0\}_{g \in SO(3)}$ onto S^2. c) Consider $(0,\ 0,\ 0) \in \mathbb{R}^3$. What is the homogeneous space containing $(0,\ 0,\ 0)$ for the usual action of $SO(3)$ on $\mathbb{R}^3$? What is the corresponding little group?

52) For the standard action of $SO(2,1)$ on $\mathbb{R}^3$, classify all the orbits and their little groups.

53) In lecture 34, the little group $G_{\hat{p}}$ for a massless particle is claimed to be $\bar{\mathcal{E}}_2$. Prove this statement. Find the UIRR's of $\bar{\mathcal{E}}_2$. For each IRR, find the eigenvectors and corresponding eigenvalues of the three generators of its Lie algebra.

CONTRACTIONS AND LIE ALGEBRAS

54) The following commutation relations define a three-parameter family of three-dimensional Lie algebras

$$[x_i, x_j] = \epsilon_{ijk} a_k x_k, \qquad i, j, k \in \{1, 2, 3\}.$$

(a) Verify that the Jacobi's identity is satisfied for any choice $a_1, a_2, a_3 \in \mathbb{R}$.
(b) Show that these algebras are the Lie algebras of unimodular Lie groups.
(c) When $a_1^2 + a_2^2 + a_3^2 \neq 0$, exhibit a procedure which contracts any one of these algebras to the Heisenberg-Weyl algebra.

HOPF ALGEBRAS

55) Consider a Hopf algebra H generated by a, b and the identity element 1 with relations $a^2 = 1$, $b^2 = 0$, $ab = -ba$. The coproduct $\triangle$ is given by $\triangle(a) = a \otimes a$, $\triangle(b) = b \otimes 1 + a \otimes b$, $\triangle(1) = 1 \otimes 1$. Derive $\epsilon(a)$, $\epsilon(b)$, $S(a)$, $S(b)$ where ϵ, and S are the counit and the antipode of H respectively.

56) Previous Exercise (55) can be generalized in the following way: Let $n \geq 2$ be an integer, r a primitive n-th root of unity. The algebra $H_{n^2}(r)$ is defined by the generators a, b with relations $a^n = 1$, $b^n = 0$, $a^{n-1}ba = rb$. The coproduct is given by $\triangle(a) = a \otimes a$, $\triangle(b) = a \otimes b + b \otimes 1$. This bialgebra has dimension n^2 and a basis given by $\{a^k b^j \mid 0 \geq k, j \geq n - 1\}$. Show that it is a Hopf algebra by determining $\epsilon(a)$, $\epsilon(b)$, $S(a)$, $S(b)$. Note: r is an n-th root of unity if $r^n = 1$, and it is primitive if in addition $r^k \neq 1$ for all $k = 1, 2, \cdots, n - 1$.

57) Let $\mathcal{A}$ be the algebra generated by an invertible element a and an element b such that $b^n = 0$ and $ab = rba$, where r is a primitive $2n$-th root of unity. Show that it is a Hopf algebra with the coproduct and counit defined by $\triangle(a) = a \otimes a$, $\triangle(b) = a \otimes b + b \otimes a^{-1}$, $\epsilon(a) = 1$, $\epsilon(b) = 0$ and determine the antipode S.

58) Let G be a connected Lie group whose Lie algebra $\mathcal{G}$ is spanned by $\{L_\alpha\}$. The universal enveloping algebra $U(\mathcal{G})$ is defined to be the algebra generated by L_α's and the identity 1 with relations given by the Lie brackets. Defining the coproduct $\triangle$, the counit ϵ and the antipode S as

$$\triangle(L_\alpha) = 1 \otimes L_\alpha + L_\alpha \otimes 1, \quad \triangle(1) = 1 \otimes 1, \quad \epsilon(L_\alpha) = 0, \quad \epsilon(1) = 1,$$
$$S(L_\alpha) = -L_\alpha, \quad S(1) = 1,$$

we can show that $U(\mathcal{G})$ becomes a Hopf algebra. Using the fact that every group element $g \in G$ can be expressed as $g = \exp(\sum_\alpha c_\alpha L_\alpha)$ show that $\triangle(g) = g \otimes g$, $\epsilon(g) = 1$, $S(g) = g^{-1}$.

59) Let H be a Hopf algebra. A twist F is an element of $H \otimes H$ that is invertible and satisfies

$$(F \otimes id)(\triangle \otimes id)F = (id \otimes F)(id \otimes \triangle)F, \quad (\epsilon \otimes id)F = (id \otimes \epsilon)F = 1.$$

Show that $\triangle_F$ defined by $\triangle_F(h) = F^{-1}\triangle(h)F$ satisfies the coassociativity condition $(\triangle_F \otimes id)\triangle_F = (id \otimes \triangle_F)\triangle_F$.

60) The dual H^* of a Hopf algebra H consists of all linear functions f from H to complex numbers $\mathbb{C}$. H^* is also canonically a Hopf algebra with product m^* and coproduct $\triangle^*$ being determined by duality from the coproduct $\triangle$ and product m of H. For example,

$$[m^*(f_1 \otimes f_2)](h) := (f_1 \otimes f_2)(\triangle(h)).$$

When G is a finite group and $\mathbb{C}G$ its Hopf algebra, what are m^* and $\triangle^*$ for its dual $\mathbb{C}G^*$? Define ϵ^* and S^* by $\epsilon^* f = f(e)$, $(S^* f)(g) = f(S(g))$, $g \in G$. Show that ϵ^* and S^* are counit and antipode for $\mathbb{C}G^*$.

61) A Hopf algebra H acts on itself on left or right thus:

$$h' \rhd_L h = h'h, \quad h \lhd_R h' = hh'.$$

These actions induce two dual actions on H^*:

$$(h'^{-1} \rhd_L f)(h) = f(h' \rhd_L h), \quad (f \lhd_R h'^{-1})(h) = f(h \lhd_R h').$$

Show that for $\mathbb{C}G^*$, these give the left and right-regular representations of G.

62) Let G be a group. Then we have seen that we can sometimes construct different Hopf algebras H from $\mathbb{C}G$ by using different coproducts $\triangle$.

Let $\mathcal{A}$ be an associative algebra with multiplication map m:

$$m : a \otimes b \rightarrow m(a \otimes b) \in \mathcal{A}, \quad a, b \in \mathcal{A}.$$

Let $\mathcal{A}$ also carry an action of G and hence of $\mathbb{C}G$.

We say that $\mathcal{A}$ is a Hopf module for the Hopf algebra H ($= \mathbb{C}G$ as a set) with coproduct $\triangle$ if

$$m\left([\triangle(g)a \otimes b]\right) = gm(a \otimes b).$$

Now the Moyal plane $\mathcal{A}_\theta(M_4)$ is the algebra of smooth functions on M_4 with the multiplication map m_θ where

$$m_\theta(f_1 \otimes f_2) = m_0(F_\theta^{-1} f_1 \otimes f_2).$$

Here m_0 is pointwise multiplication,

$$m_0(f_1 \otimes f_2)(x) = f_1(x)f_2(x)$$

and F_θ is the Drinfel'd twist defined in the text.

Let $H_\theta \mathcal{P}$ be the Hopf algebra for the Poincaré group algebra $\mathbb{C}\mathcal{P}$ and the coproduct $\triangle_\theta$ as in the text.

Show that $\mathcal{A}_\theta(M_4)$ is a Hopf module algebra for $H_\theta \mathcal{P}$.

63) Consider the algebra $\mathcal{A}$ with generators $a, x_1, x_2, \cdots, x_n$ subject to the relations $a^2 = 1$, $x_k^2 = 0 \; \forall k$, $x_k a = -a x_k$, $x_j x_k = -x_k x_j$. Define coproduct by $\triangle(a) = a \otimes a$, $\triangle(x_k) = a \otimes x_k + x_k \otimes 1$. Show that $\mathcal{A}$ then becomes a Hopf algebra, and determine ϵ and S.

64) If $\mu \colon P \times P \to P$ is an associative product on the smooth manifold P, the vector space $\mathcal{F}(P, \mathbb{C})$, of smooth complex valued functions, is a bialgebra with respect to the pointwise-product and the coproduct given by $\triangle \colon \mathcal{F}(P) \to \mathcal{F}(P \times P) \cong \mathcal{F}(P) \times \mathcal{F}(P)$ with $(\triangle f)(p_1, p_2) = f(\mu(p_1, p_2))$. What additional requirements we need on (P, μ) to make $\mathcal{F}(P)$ into a Hopf algebra? Notice that this will be the case if P is a Lie group and μ the corresponding multiplication.

Bibliography

This bibliography makes no pretense to completeness.

GENERAL TREATISES ON GROUP THEORY

1) H. Bacry, *Lectures on Group Theory and Particle Theory* (Gordon and Breach, 1977).

2) H. Boerner, *Representations of Groups with Special Consideration for the Needs of Modern Physics* (North Holland, 1970).

3) W. Fulton, J. Harris, *Representation Theory, A First Course* (Springer-Verlag, New York, 1991).

4) M. Hamermesh, *Group Theory and Its Applications to Physical Problems* (Addison-Wesley, 1962).

5) TH. Kahan, Ed., *The Theory of Groups in Classical and Quantum Physics*, Vol.I (American Elsevier, 1966).

6) A. Kirillov, *Elements of the Theory of Representations* (Springer-Verlag, Berlin-New York, 1976).

7) M. Naimark, A. Stern, *Theory of Group Representations* (Springer-Verlag, New York, 1982).

8) L. Pontrjagin, *Topological Groups* (Princeton University Press, 1958).

9) S. Sternberg, *Group Theory and Physics* (Cambridge University Press, Cambridge, 1995).

10) H. Weyl, *The Classical Groups* (Princeton University Press, 1946).

FINITE GROUPS

References 1,2,4,5,10.

11) W. Ledermann, *Introduction to the Theory of Finite Groups* (Oliver and Boyd, Edinberg, 1964).

12) F. Murnaghan, *The Theory of Group Representations* (Dover,

1963).

13) H. Weyl, *Group Theory and Quantum Mechanics* (Dover, 1950).

14) E. P. Wigner, *Group Theory and its Applications to the Quantum Mechanics of Atomic Spectra* (Academic Press, 1959).

LIE GROUPS

References 1 to 10.

15) R. Carter, G. Segal, I. MacDonald, *Lectures on Lie Groups and Lie Algebras* (Cambridge University Press, Cambridge, 1995).

16) H. Georgi, *Lie Algebras in Particle Physics* (Benhamin/Cummings. 1982).

17) R. Gilmore, *Lie Groups, Lie Algebras and Some of Their Applications* (Wiley, 1974).

18) R. W. Goodman, N. R. Wallach, *Representations and Invariants of the Classical Groups* (Cambridge University Press, Cambridge, 1998).

19) M. Gourdin, *Basics of Lie Groups* (Éditions Frontiérs, 1982).

20) R. Hermann, *Lie Groups for Physicists* (Benjamin, 1966).

21) R. Hermann, *Lie Algebras and Quantum Mechanics* (Benjamin, 1970).

22) W. Miller, *Lie Theory and Special Functions* (Academic Press, 1968).

23) C. Procesi, *Lie Groups, An Approach Through Invariants And Representations* (Springer-Verlag, New York, 2007).

24) G. Racah, *Group Theory and Spectroscopy*, Springer Tracts in Modern Physics, Vol.37 (1965).

25) J. Talman, E. P. Wigner, *Special Functions, a Group Theoretical Approach* (Benjamin, 1968).

26) V. S. Varadarajan, *Lie Groups, Lie Algebras, and Their Representations* (Springer, New York, 1984).

27) N. Vilenkin, *Special Functions and the Theory of Group Representations* (American Mathematical Society Translations, 1968).

28) H. Weyl, *The Structure and Representations of Continuous Groups* (Princeton University Press, 1935).

BIANCHI CLASSIFICATION

29) J. M. Gracia-Bondia, F. Lizzi, G. Marmo, P. Vitale, *Infinitely many star products to play with*, J. High Energy Phys. 04(2002)026.

30) S. Capozziello, G. Marmo, C. Rubano, P. Scudellaro, *Nöther Symmetries in Bianchi Universes*, Internat. J. Modern Phys. D6(1997), 491-503.

THE ROTATION GROUP AND ANGULAR MOMENTUM

References 1,2,4,5,13,14,16,22,24,25,27.

31) L. C. Biedenharn, H. Van Dam, *Quantum Theory of Angular Momentum* (Academic Press, 1965).

32) D. N. Brink, G. R. Satchler, *Angular Momentum* (Clarendon, 1968).

33) A. R. Edmonds, *Angular Momentum in Quantum Mechanics* (Princeton University Press, 1968).

34) U. Fano, G. Racah, *Irreducible Tensorial Sets* (Academic press, 1959).

35) I. M. Gelfand, R. A. Minlos, Z. Ya. Shapiro, *Representations of the Rotation and the Lorentz Groups and their Applications* (MacMillan, 1963).

36) C. Muller, *Lectures on the Theory of Spherical Harmonics* (Springer Verlag, 1966).

37) M. E. Rose, *Elementary Theory of Angular Momentum* (Wiley, 1957).

THE LORENTZ GROUP

References 1, 2, 35.

38) I. M. Gelfand, M. I. Graev, N. Y. Vilenkin, *Generalized Functions*, Vol.5 (Academic Press, 1966).

39) M. A. Naimark, *Linear Representations of the Lorentz Group* (Pergamon, 1964).

THE POINCARÉ GROUP

Reference 1.

40) P. Moussa, R. Stora, in W. E. Britten and A. O. Barut, Eds., *Lectures in Theoretical Physics* (University of Colorado Press, 1964).

41) E. P. Wigner, in F. J. Dyson, Ed., *Symmetry Groups in Nuclear and Particle Physics* (Benjamin, 1966).

42) E. P. Wigner, in F. Gürsey, Ed., *Group Theoretical Concepts and*

Methods in Elementary Particle Physics (Gordon and Breach, 1964).

CONTRACTIONS

43) J. F. Carinena, G. Marmo, J. Grabowski, *Contractions: Nijenhuis and Saletan tensors for general algebraic structures*, J. Phys. A(2001), 3769-3789.

HOPF ALGEBRAS

44) V. Chari, A. Pressley, *A Guide to Quantum Groups* (Cambridge University Press, 1994).

45) S. Dascalescu, C. Nastasescu, S. Raianu, *Hopf Algebra, An Introduction* (M. Dekker, New York, 2001).

46) G. Mack, V. Schomerus, *New Symmetry Principles in Quantum Field Theory* (NATO Science Series :B:)(Springer-Verlag, Berlin, 1992); Nucl. Phys. B370 (1992), 185; DESY 92-053 (1992).

47) J. W. Milnor, J. C. Moore, *On the structure of Hopf Algebras*, Annals of Math. 81(1965), 211-269.

48) Shahn Majid, *Foundations of Quantum Group Theory* (Cambridge University Press, 1995).

49) Moss E. Sweedler, *Hopf Algebra* (W. A. Benjamin, New York, 1969).

PROBLEMS IN GROUP THEORY

50) E. S. Lyapin, A. Ya. Ajzenshtat, M. M. Lesokhin, *Exercises in Group Theory* (Plenum Press, 1972).

Index